2-D Geometry

Turtle Paths

Grade 3

Also appropriate for Grade 4

Douglas H. Clements
Michael T. Battista
Joan Akers
Virginia Woolley
Julie Sarama Meredith
Sue McMillen

Developed at TERC, Cambridge, Massachusetts

Dale Seymour Publications®
Menlo Park, California

The *Investigations* curriculum was developed at TERC (formerly
Technical Education Research Centers) in collaboration with Kent State
University and the State University of New York at Buffalo. The work
was supported in part by National Science Foundation Grant No. ESI-
9050210. TERC is a nonprofit company working to improve mathematics
and science education. TERC is located at 2067 Massachusetts Avenue,
Cambridge, MA 02140.

This project was supported, in part,
by the
National Science Foundation
Opinions expressed are those of the authors
and not necessarily those of the Foundation

Managing Editor: Catherine Anderson
Series Editor: Beverly Cory
Revision Team: Laura Marshall Alavosus, Ellen Harding, Patty Green
Holubar, Suzanne Knott, Beverly Hersh Lozoff
ESL Consultant: Nancy Sokol Green
Production/Manufacturing Director: Janet Yearian
Production/Manufacturing Coordinator: Barbara Atmore
Design Manager: Jeff Kelly
Design: Don Taka
Illustrations: Barbara Epstein-Eagle, Hollis Burkhart
Cover: Bay Graphics
Composition: Archetype Book Composition

This book is published by Dale Seymour Publications®, an imprint of
Addison Wesley Longman, Inc.

Dale Seymour Publications
2725 Sand Hill Road
Menlo Park, CA 94025
Customer Service: 800-872-1100

Order number DS43848
ISBN 1-57232-701-4
1 2 3 4 5 6 7 8 9 10-ML-01 00 99 98 97

Printed on Recycled Paper

TERC

INVESTIGATIONS IN NUMBER, DATA, AND SPACE®

Principal Investigator   Susan Jo Russell

Co-Principal Investigator   Cornelia C. Tierney

Director of Research and Evaluation   Jan Mokros

Curriculum Development
Joan Akers
Michael T. Battista
Mary Berle-Carman
Douglas H. Clements
Karen Economopoulos
Ricardo Nemirovsky
Andee Rubin
Susan Jo Russell
Cornelia C. Tierney
Amy Shulman Weinberg

Evaluation and Assessment
Mary Berle-Carman
Abouali Farmanfarmaian
Jan Mokros
Mark Ogonowski
Amy Shulman Weinberg
Tracey Wright
Lisa Yaffee

Teacher Support
Rebecca B. Corwin
Karen Economopoulos
Tracey Wright
Lisa Yaffee

Technology Development
Michael T. Battista
Douglas H. Clements
Julie Sarama Meredith
Andee Rubin

Video Production
David A. Smith

Administration and Production
Amy Catlin
Amy Taber

Cooperating Classrooms
for This Unit
Sharon Cacicia
Janice Wagner
Starpoint Central School
Lockport, NY

Sue Combs
Joyce Golibersuch
Susan Hunt
Buffalo Public School System
Buffalo, NY

Diane McCarthy
Niagara-Wheatfield Central School
Sanborn, NY

Virginia M. Micciche
Cambridge Public Schools
Cambridge, MA

Jeanne Wall
Arlington Public Schools
Arlington, MA

Katie Bloomfield
Robert A. Dihlmann
Shutesbury Elementary
Shutesbury, MA

Consultants and Advisors
Elizabeth Badger
Deborah Lowenberg Ball
Marilyn Burns
Ann Grady
Joanne M. Gurry
James J. Kaput
Steven Leinwand
Mary M. Lindquist
David S. Moore
John Olive

Leslie P. Steffe
Peter Sullivan
Grayson Wheatley
Virginia Woolley
Anne Zarinnia

Graduate Assistants
Kent State University
Joanne Caniglia
Pam DeLong
Carol King

State University of New York at Buffalo
Rosa Gonzalez
Sue McMillen
Julie Sarama Meredith
Sudha Swaminathan

Revisions and Home Materials
Cathy Miles Grant
Marlene Kliman
Margaret McGaffigan
Megan Murray
Kim O'Neil
Andee Rubin
Susan Jo Russell
Lisa Seyferth
Myriam Steinback
Judy Storeygard
Anna Suarez
Cornelia Tierney
Carol Walker
Tracey Wright

CONTENTS

TEACHER NOTES

WHERE TO START

The first-time user of *Turtle Paths* should read the following:

When you next teach this same unit, you can begin to read more of the background. Each time you present the unit, you will learn more about how your students understand the mathematical ideas.

Investigations in Number, Data, and Space® is a K–5 mathematics curriculum with four major goals:

- to offer students meaningful mathematical problems
- to emphasize depth in mathematical thinking rather than superficial exposure to a series of fragmented topics
- to communicate mathematics content and pedagogy to teachers
- to substantially expand the pool of mathematically literate students

The *Investigations* curriculum embodies a new approach based on years of research about how children learn mathematics. Each grade level consists of a set of separate units, each offering 2–8 weeks of work. These units of study are presented through investigations that involve students in the exploration of major mathematical ideas.

Approaching the mathematics content through investigations helps students develop flexibility and confidence in approaching problems, fluency in using mathematical skills and tools to solve problems, and proficiency in evaluating their solutions. Students also build a repertoire of ways to communicate about their mathematical thinking, while their enjoyment and appreciation of mathematics grows.

The investigations are carefully designed to invite all students into mathematics—girls and boys, members of diverse cultural, ethnic, and language groups, and students with different strengths and interests. Problem contexts often call on students to share experiences from their family, culture, or community. The curriculum eliminates barriers—such as work in isolation from peers, or emphasis on speed and memorization—that exclude some students from participating successfully in mathematics. The following aspects of the curriculum ensure that all students are included in significant mathematics learning:

- Students spend time exploring problems in depth.
- They find more than one solution to many of the problems they work on.

- They invent their own strategies and approaches, rather than relying on memorized procedures.
- They choose from a variety of concrete materials and appropriate technology, including calculators, as a natural part of their everyday mathematical work.
- They express their mathematical thinking through drawing, writing, and talking.
- They work in a variety of groupings—as a whole class, individually, in pairs, and in small groups.
- They move around the classroom as they explore the mathematics in their environment and talk with their peers.

While reading and other language activities are typically given a great deal of time and emphasis in elementary classrooms, mathematics often does not get the time it needs. If students are to experience mathematics in depth, they must have enough time to become engaged in real mathematical problems. We believe that a minimum of five hours of mathematics classroom time a week—about an hour a day—is critical at the elementary level. The plan and pacing of the *Investigations* curriculum is based on that belief.

We explain more about the pedagogy and principles that underlie these investigations in Teacher Notes throughout the units. For correlations of the curriculum to the NCTM Standards and further help in using this research-based program for teaching mathematics, see the following books:

- *Implementing the* Investigations in Number, Data, and Space® *Curriculum*
- *Beyond Arithmetic: Changing Mathematics in the Elementary Classroom* by Jan Mokros, Susan Jo Russell, and Karen Economopoulos

This book is one of the curriculum units for *Investigations in Number, Data, and Space.* In addition to providing part of a complete mathematics curriculum for your students, this unit offers information to support your own professional development. You, the teacher, are the person who will make this curriculum come alive in the classroom; the book for each unit is your main support system.

Although the curriculum does not include student textbooks, reproducible sheets for student work are provided in the unit and are also available as Student Activity Booklets. Students work actively with objects and experiences in their own environment and with a variety of manipulative materials and technology, rather than with a book of instruction and problems. We strongly recommend use of the overhead projector as a way to present problems, to focus group discussion, and to help students share ideas and strategies.

Ultimately, every teacher will use these investigations in ways that make sense for his or her particular style, the particular group of students, and the constraints and supports of a particular school environment. Each unit offers information and guidance for a wide variety of situations, drawn from our collaborations with many teachers and students over many years. Our goal in this book is to help you, a professional educator, implement this curriculum in a way that will give all your students access to mathematical power.

Investigation Format

The opening two pages of each investigation help you get ready for the work that follows.

What Happens This gives a synopsis of each session or block of sessions.

Mathematical Emphasis This lists the most important ideas and processes students will encounter in this investigation.

What to Plan Ahead of Time These lists alert you to materials to gather, sheets to duplicate, transparencies to make, and anything else you need to do before starting.

INVESTIGATION 1

Paths and Lengths of Paths

What Happens

Session 1: Walking Paths Session 1 begins with two whole-class activities. Students walk on masking-tape paths on the floor, describing the movements they make. Students create paths for others by giving them movement commands (such as "forward 5 steps, left turn 90, forward 2 steps"). The session continues with the introduction of two or three Off-Computer Choices (Off-Computer Choice 3 can be introduced in Session 2). Students work in pairs to make paths on dot paper and to count steps in a maze to find different paths, such as one that is 14 steps in length and has 2 corners.

Session 2: Commanding the Turtle In Session 2, students are introduced to *Geo-Logo* and the first On-Computer Activity, Get the Toys. If they were not introduced to Off-Computer Choice 3, Maze Game, in Session 1, they are introduced to that activity during this session. Then students are divided into two groups—one working on the computers and the other working at their seats. The groups are switched halfway through the work period. Within each group, students work in pairs to find and describe paths. Students working on the computer give commands to the *Geo-Logo* turtle to fetch toys located in various positions on floor maps. Students working at their seats continue the Off-Computer Choices introduced in Sessions 1 and 2.

Sessions 3 and 4: Mazes and Maps Session 3 begins with a review of *Geo-Logo* and the On-Computer Activity, Get the Toys. It continues with an explanation of how to use *Geo-Logo* tools to "teach" and "run" procedures students have created. Session 3 continues with the class being divided into two groups. Working in pairs, one

group continues to work on the On-Computer Activity, Get the Toys, and the other group continues with the Off-Computer Choices introduced in Sessions 1 and 2.

Session 4 begins after all students have had the opportunity to work at the computer. Students share their strategies and solutions to how they got the toys. Session 4 continues with the class being divided into two groups. Working in pairs, one group continues to work on the On-Computer Activity, Get the Toys, and the other continues to work on the Off-Computer Choices introduced in Sessions 1 and 2.

Mathematical Emphasis

■ Understanding paths as representations or records of movement

■ Describing paths with mathematical language (for example, "closed," "corner")

■ Finding several paths that meet geometric constraints (for example, finding several ways to create a path that is 30 steps long with 2 corners)

■ Using *Geo-Logo* commands to construct paths and describe their properties

■ Applying mathematical processes such as addition, subtraction, estimation, and "undoing" to paths

---

INVESTIGATION 1

What to Plan Ahead of Time

Materials

■ Computers—Macintosh II or above, with 4 MB of internal memory (RAM) and Apple System Software 7.0 or later. Maximum: 1 for every 2 students. Minimum: 1 for every 4–6 students. It is possible to modify the unit for fewer computers. (See Managing the Computer Activities in This Unit, p. I-21).

■ Apple Macintosh disk, *Geo-Logo™*, for *Turtle Paths* (Sessions 2–4)

■ A large-screen monitor on one computer for whole-class viewing (recommended)

■ Masking tape (Session 1)

■ Small counters or cubes: 10–15 (Sessions 1–4)

■ Number cubes: 8–10 (Session 1–4)

■ Colored pencils or crayons for each group of students (Sessions 1–4)

■ Overhead projector

Other Preparation

■ Duplicate student sheets and teaching resources (located at the end of the unit) in the following quantities. If you have Student Activity Booklets, copy only the items marked with an asterisk, including any extra materials and transparencies needed.

For Sessions 1–4

Student Sheet 1, Maze Paths (p. 133): several copies per student, and 1 overhead transparency*

Student Sheet 2, Maze Paths Challenges (p. 134): 1 per student, and 1 overhead transparency*

Student Sheet 3, Paths at Home (p. 135): 1 per student (homework)

Geo-Logo User Sheet* (p. 139): 1 per computer

Dot Paper (p. 140): 1 sheet per student

Family letter* (p. 132): 1 per student. Remember to sign it before copying.

For Sessions 2–4

Student Sheet 4, Floors 1 and 2 (p. 136): several per student, and 1 overhead transparency*

Student Sheet 5, Floor 3 (p. 137): 1 per student

Student Sheet 6, Many Possible Paths (p. 138): 1 per student (homework)

■ Use the disk for *Turtle Paths* to install *Geo-Logo* on each computer. (See pp. 128–129 in the *Geo-Logo* Teacher Tutorial.)

■ Work through the following sections of the *Geo-Logo* Teacher Tutorial.

    Overview (p. 89)

    Getting Started with *Geo-Logo* (p. 91)

    Get the Toys

        How to Start an Activity (p. 93)

        How to Play Get the Toys on Floor 1 (p. 93)

        How to Save Your Game or Work (p. 97)

        How to Finish an Activity (p. 98)

Continued on next page

Sessions Within an investigation, the activities are organized by class session, a session being at least a one-hour math class. Sessions are numbered consecutively through an investigation. Often several sessions are grouped together, presenting a block of activities with a single major focus.

When you find a block of sessions presented together—for example, Sessions 1, 2, and 3—read through the entire block first to understand the overall flow and sequence of the activities. Make some preliminary decisions about how you will divide the activities into three sessions for your class, based on what you know about your students. You may need to modify your initial plans as you progress through the activities, and you may want to make notes in the margins of the pages as reminders for the next time you use the unit.

Be sure to read the Session Follow-Up section at the end of the session block to see what homework assignments and extensions are suggested as you make your initial plans.

While you may be used to a curriculum that tells you exactly what each class session should cover, we have found that the teacher is in a better position to make these decisions. Each unit is flexible and may be handled somewhat differently by every teacher. While we provide guidance for how many sessions a particular group of activities is likely to need, we want you to be active in determining an appropriate pace and the best transition points for your class. It is not unusual for a teacher to spend more or less time than is proposed for the activities.

Ten-Minute Math At the beginning of some sessions, you will find Ten-Minute Math activities. These are designed to be used in tandem with the investigations, but not during the math hour. Rather, we hope you will do them whenever you have a spare 10 minutes—maybe before lunch or recess, or at the end of the day.

Ten-Minute Math offers practice in key concepts, but not always those being covered in the unit. For example, in a unit on using data, Ten-Minute Math might revisit geometric activities done earlier in the year. Complete directions for the suggested activities are included at the end of each unit.

Session 2

Commanding the Turtle

Materials

- Choice Time materials from Session 1
- Student Sheet 4 (several per student)
- Transparency of Student Sheet 4
- Geo-Logo User Sheet (1 per computer)
- Computers with Geo-Logo installed
- Overhead projector

What Happens

In Session 2, students are introduced to Geo-Logo and the first On-Computer Activity, Get the Toys. If they were not introduced to Off-Computer Choice 3, Maze Game, in Session 1, they are introduced to that activity during this session. Then students are divided into two groups—one working at the computers and the other working at their seats. The groups are switched halfway through the work period. Within each group, students work in pairs to find and describe paths. Students working at the computer give commands to the Logo turtle to fetch toys located in various positions on floor maps. Students working at their seats continue the Off-Computer Choices introduced in Sessions 1 and 2. Their work focuses on:

- finding alternative solutions to path problems that pose specific geometric constraints
- using Geo-Logo commands to construct paths on a map to reach a given location
- applying mathematical processes such as addition, subtraction, estimation, and "undoing" to Geo-Logo commands to figure out the most efficient way to move on a map
- combining commands efficiently

The chart below shows how students work during this session.

Session 2		
Whole Class 30 min	**Introducing Work Procedures** **On-Computer Activity: Get the Toys** (Introduction)	
Two Groups (working in pairs) 15 min	Group A **On-Computer Activity** ■ Get the Toys	Group B **Off-Computer Choices** ■ Dot Paths ■ Maze Paths ■ Maze Game
Switch 15 min	**Off-Computer Choices**	**On-Computer Activity**

Activities The activities include pair and small-group work, individual tasks, and whole-class discussions. In any case, students are seated together, talking and sharing ideas during all work times. Students most often work cooperatively, although each student may record work individually.

Choice Time In some units, some sessions are structured with activity choices. In these cases, students may work simultaneously on different activities focused on the same mathematical ideas. Students choose which activities they want to do, and they cycle through them.

You will need to decide how to set up and introduce these activities and how to let students make their choices. Some teachers present them as station activities, in different parts of the room. Some list the choices on the board as reminders or have students keep their own lists.

Extensions Sometimes in Session Follow-Up, you will find suggested extension activities. These are opportunities for some or all students to explore a topic in greater depth or in a different context. They are not designed for "fast" students; mathematics is a multifaceted discipline, and different students will want to go further in different investigations. Look for and encourage the sparks of interest and enthusiasm you see in your students, and use the extensions to help them pursue these interests.

Excursions Some of the *Investigations* units include excursions—blocks of activities that could be omitted without harming the integrity of the unit. This is one way of dealing with the great depth and variety of elementary mathematics— much more than a class has time to explore in any one year. Excursions give you the flexibility to make different choices from year to year, doing the excursion in one unit this time, and next year trying another excursion.

Tips for the Linguistically Diverse Classroom At strategic points in each unit, you will find concrete suggestions for simple modifications of the teaching strategies to encourage the participation of all students. Many of these tips offer alternative ways to elicit critical thinking from students at varying levels of English proficiency, as well as from other students who find it difficult to verbalize their thinking.

The tips are supported by suggestions for specific vocabulary work to help ensure that all students can participate fully in the investigations. The Preview for the Linguistically Diverse Classroom (p. I-20) lists important words that are assumed as part of the working vocabulary of the unit. Second-language learners will need to become familiar with these words in order to understand the problems and activities they will be doing. These terms can be incorporated into students' second-language work before or during the unit. Activities that can be used to present the words are found in the appendix, Vocabulary Support for Second-Language Learners (p. 85). In addition, ideas for making connections to students' language and cultures, included on the Preview page, help the class explore the unit's concepts from a multicultural perspective.

Materials

A complete list of the materials needed for teaching this unit is found on p. I-16. Some of these materials are available in kits for the *Investigations* curriculum. Individual items can also be purchased from school supply dealers.

Classroom Materials In an active mathematics classroom, certain basic materials should be available at all times: interlocking cubes, pencils, unlined paper, graph paper, calculators, things to count with, and measuring tools. Some activities in this curriculum require scissors and glue sticks or tape. Stick-on notes and large paper are also useful materials throughout.

So that students can independently get what they need at any time, they should know where these materials are kept, how they are stored, and how they are to be returned to the storage area. For example, interlocking cubes are best stored in towers of ten; then, whatever the activity, they should be returned to storage in groups of ten at the end of the hour. You'll find that establishing such routines at the beginning of the year is well worth the time and effort.

Technology Calculators are used throughout *Investigations.* Many of the units recommend that you have at least one calculator for each pair. You will find calculator activities, plus Teacher Notes discussing this important mathematical tool, in an early unit at each grade level. It is assumed that calculators will be readily available for student use.

Computer activities at grade 3 use two software programs that were developed especially for the *Investigations* curriculum. *Tumbling Tetrominoes* is introduced in the 2-D Geometry unit, *Flips, Turns, and Area.* This game emphasizes ideas about area and about geometric motions (slides, flips, and turns). The program *Geo-Logo*™ is introduced in a second 2-D Geometry unit, *Turtle Paths,* where students use it to explore geometric shapes.

How you use the computer activities depends on the number of computers you have available. Suggestions are offered in the geometry units for how to organize different types of computer environments.

Children's Literature Each unit offers a list of suggested children's literature (p. I-16) that can be used to support the mathematical ideas in the unit. Sometimes an activity is based on a specific children's book, with suggestions for substitutions where practical. While such activities can be adapted and taught without the book, the literature offers a rich introduction and should be used whenever possible.

Student Sheets and Teaching Resources Student recording sheets and other teaching tools needed for both class and homework are provided as reproducible blackline masters at the end of each unit. They are also available as Student Activity Booklets. These booklets contain all the sheets each student will need for individual work, freeing you from extensive copying (although you may need or want to copy the occasional teaching resource on transparency film or card stock, or make extra copies of a student sheet).

We think it's important that students find their own ways of organizing and recording their work. They need to learn how to explain their thinking with both drawings and written words, and how to organize their results so someone else can under-

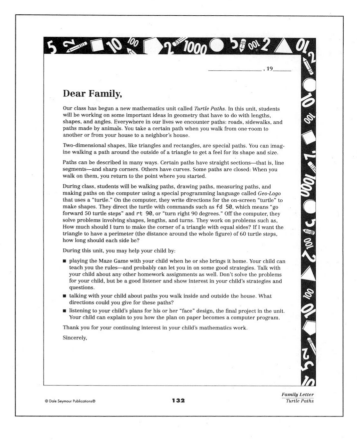

stand them. For this reason, we deliberately do not provide student sheets for every activity. Regardless of the form in which students do their work, we recommend that they keep a mathematics notebook or folder so that their work is always available for reference.

Homework In *Investigations,* homework is an extension of classroom work. Sometimes it offers review and practice of work done in class, sometimes preparation for upcoming activities, and sometimes numerical practice that revisits work in earlier units. Homework plays a role both in supporting students' learning and in helping inform families about the ways in which students in this curriculum work with mathematical ideas.

Depending on your school's homework policies and your own judgment, you may want to assign more homework than is suggested in the units. For this purpose you might use the practice pages, included as blackline masters at the end of this unit, to give students additional work with numbers.

For some homework assignments, you will want to adapt the activity to meet the needs of a variety of students in your class: those with special needs, those ready for more challenge, and second-language learners. You might change the numbers in a problem, make the activity more or less complex, or go through a sample activity with those who need extra help. You can modify any student sheet for either homework or class use. In particular, making numbers in a problem smaller or larger can make the same basic activity appropriate for a wider range of students.

Another issue to consider is how to handle the homework that students bring back to class—how to recognize the work they have done at home without spending too much time on it. Some teachers hold a short group discussion of different approaches to the assignment; others ask students to share and discuss their work with a neighbor, or post the homework around the room and give students time to tour it briefly. If you want to keep track of homework students bring in, be sure it ends up in a designated place.

Investigations at Home It is a good idea to make your policy on homework explicit to both students and their families when you begin teaching with *Investigations*. How frequently will you be assigning homework? When do you expect homework to be completed and brought back to school? What are your goals in assigning homework? How independent should families expect their children to be? What should the parent's or guardian's role be? The more explicit you can be about your expectations, the better the homework experience will be for everyone.

Investigations at Home (a booklet available separately for each unit, to send home with students) gives you a way to communicate with families about the work students are doing in class. This booklet includes a brief description of every session, a list of the mathematics content emphasized in each investigation, and a discussion of each homework assignment to help families more effectively support their children. Whether or not you are using the *Investigations* at Home booklets, we expect you to make your own choices about home-

work assignments. Feel free to omit any and to add extra ones you think are appropriate.

Family Letter A letter that you can send home to students' families is included with the blackline masters for each unit. Families need to be informed about the mathematics work in your classroom; they should be encouraged to participate in and support their children's work. A reminder to send home the letter for each unit appears in one of the early investigations. These letters are also available separately in Spanish, Vietnamese, Cantonese, Hmong, and Cambodian.

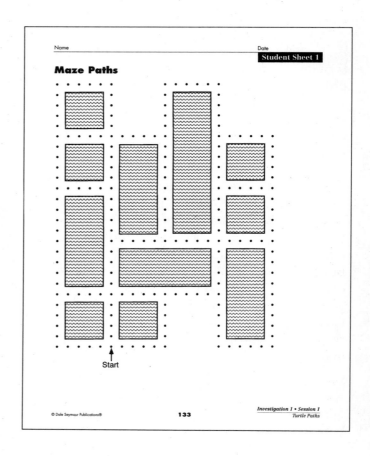

Name _____ Date _____

Maze Paths

Start

133 *Investigation 1 • Session 1*
Turtle Paths

Help for You, the Teacher

Because we believe strongly that a new curriculum must help teachers think in new ways about mathematics and about their students' mathematical thinking processes, we have included a great deal of material to help you learn more about both.

About the Mathematics in This Unit This introductory section (p. I-17) summarizes the critical information about the mathematics you will be teaching. It describes the unit's central mathematical ideas and how students will encounter them through the unit's activities.

Teacher Notes These reference notes provide practical information about the mathematics you are teaching and about our experience with how students learn. Many of the notes were written in response to actual questions from teachers, or to discuss important things we saw happening in the field-test classrooms. Some teachers like to read them all before starting the unit, then review them as they come up in particular investigations.

Dialogue Boxes Sample dialogues demonstrate how students typically express their mathematical ideas, what issues and confusions arise in their thinking, and how some teachers have guided class discussions.

These dialogues are based on the extensive classroom testing of this curriculum; many are word-for-word transcriptions of recorded class discussions. They are not always easy reading; sometimes it may take some effort to unravel what the students are trying to say. But this is the value of these dialogues; they offer good clues to how your students may develop and express their approaches and strategies, helping you prepare for your own class discussions.

Where to Start You may not have time to read everything the first time you use this unit. As a first-time user, you will likely focus on understanding the activities and working them out with your students. Read completely through each investigation before starting to present it. Also read those sections listed in the Contents under the heading Where to Start (p. vi).

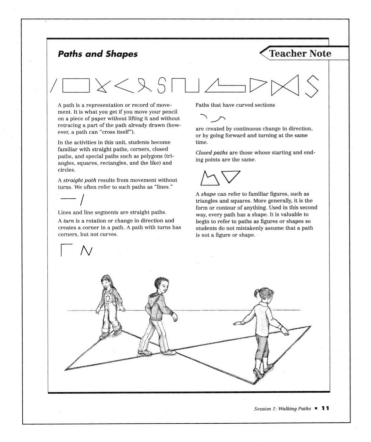

The *Investigations* curriculum incorporates the use of two forms of technology in the classroom: calculators and computers. Calculators are assumed to be standard classroom materials, available for student use in any unit. Computers are explicitly linked to one or more units at each grade level; they are used with the unit on 2-D geometry at each grade, as well as with some of the units on measuring, data, and changes.

Using Calculators

In this curriculum, calculators are considered tools for doing mathematics, similar to pattern blocks or interlocking cubes. Just as with other tools, students must learn both *how* to use calculators correctly and *when* they are appropriate to use. This knowledge is crucial for daily life, as calculators are now a standard way of handling numerical operations, both at work and at home.

Using a calculator correctly is not a simple task; it depends on a good knowledge of the four operations and of the number system, so that students can select suitable calculations and also determine what a reasonable result would be. These skills are the basis of any work with numbers, whether or not a calculator is involved.

Unfortunately, calculators are often seen as tools to check computations with, as if other methods are somehow more fallible. Students need to understand that any computational method can be used to check any other; it's just as easy to make a mistake on the calculator as it is to make a mistake on paper or with mental arithmetic. Throughout this curriculum, we encourage students to solve computation problems in more than one way in order to double-check their accuracy. We present mental arithmetic, paper-and-pencil computation, and calculators as three possible approaches.

In this curriculum we also recognize that, despite their importance, calculators are not always appropriate in mathematics instruction. Like any tools, calculators are useful for some tasks, but not for others. You will need to make decisions about when to allow students access to calculators and when to ask that they solve problems without them, so that they can concentrate on other tools and skills. At times when calculators are or are not appropriate for a particular activity, we make specific recommendations. Help your students develop their own sense of which problems they can tackle with their own reasoning and which ones might be better solved with a combination of their own reasoning and the calculator.

Managing calculators in your classroom so that they are a tool, and not a distraction, requires some planning. When calculators are first introduced, students often want to use them for everything, even problems that can be solved quite simply by other methods. However, once the novelty wears off, students are just as interested in developing their own strategies, especially when these strategies are emphasized and valued in the classroom. Over time, students will come to recognize the ease and value of solving problems mentally, with paper and pencil, or with manipulatives, while also understanding the power of the calculator to facilitate work with larger numbers.

Experience shows that if calculators are available only occasionally, students become excited and distracted when they are permitted to use them. They focus on the tool rather than on the mathematics. In order to learn when calculators are appropriate and when they are not, students must have easy access to them and use them routinely in their work.

If you have a calculator for each student, and if you think your students can accept the responsibility, you might allow them to keep their calculators with the rest of their individual materials, at least for the first few weeks of school. Alternatively, you might store them in boxes on a shelf, number each calculator, and assign a corresponding number to each student. This system can give students a sense of ownership while also helping you keep track of the calculators.

Using Computers

Students can use computers to approach and visualize mathematical situations in new ways. The computer allows students to construct and manipulate geometric shapes, see objects move according to rules they specify, and turn, flip, and repeat a pattern.

This curriculum calls for computers in units where they are a particularly effective tool for learning mathematics content. One unit on 2-D geometry at each of the grades 3–5 includes a core of activities that rely on access to computers, either in the classroom or in a lab. Other units on geometry, measurement, data, and changes include computer activities, but can be taught without them. In these units, however, students' experience is greatly enhanced by computer use.

The following list outlines the recommended use of computers in this curriculum:

Grade 1
Unit: *Survey Questions and Secret Rules*
   (Collecting and Sorting Data)
Software: Tabletop, Jr.
Source: Broderbund

Unit: *Quilt Squares and Block Towns*
   (2-D and 3-D Geometry)
Software: *Shapes*
Source: provided with the unit

Grade 2
Unit: *Mathematical Thinking at Grade 2*
   (Introduction)
Software: *Shapes*
Source: provided with the unit

Unit: *Shapes, Halves, and Symmetry*
   (Geometry and Fractions)
Software: *Shapes*
Source: provided with the unit

Unit: *How Long? How Far?* (Measuring)
Software: *Geo-Logo*
Source: provided with the unit

Grade 3
Unit: *Flips, Turns, and Area* (2-D Geometry)
Software: *Tumbling Tetrominoes*
Source: provided with the unit

Unit: *Turtle Paths* (2-D Geometry)
Software: *Geo-Logo*
Source: provided with the unit

Grade 4
Unit: *Sunken Ships and Grid Patterns*
   (2-D Geometry)
Software: *Geo-Logo*
Source: provided with the unit

Grade 5
Unit: *Picturing Polygons* (2-D Geometry)
Software: *Geo-Logo*
Source: provided with the unit

Unit: *Patterns of Change* (Tables and Graphs)
Software: *Trips*
Source: provided with the unit

Unit: *Data: Kids, Cats, and Ads* (Statistics)
Software: Tabletop, Sr.
Source: Broderbund

The software provided with the *Investigations* units uses the power of the computer to help students explore mathematical ideas and relationships that cannot be explored in the same way with physical materials. With the *Shapes* (grades 1–2) and *Tumbling Tetrominoes* (grade 3) software, students explore symmetry, pattern, rotation and reflection, area, and characteristics of 2-D shapes. With the *Geo-Logo* software (grades 3–5), students investigate rotations and reflections, coordinate geometry, the properties of 2-D shapes, and angles. The *Trips* software (grade 5) is a mathematical exploration of motion in which students run experiments and interpret data presented in graphs and tables.

We suggest that students work in pairs on the computer; this not only maximizes computer resources but also encourages students to consult, monitor, and teach one another. Generally, more than two students at one computer find it difficult to share. Managing access to computers is an issue for every classroom. The curriculum gives you explicit support for setting up a system. The units are structured on the assumption that you have enough computers for half your students to work on the machines in pairs at one time. If you do not have access to that many computers, suggestions are made for structuring class time to use the unit with five to eight computers, or even with fewer than five.

Assessment plays a critical role in teaching and learning, and it is an integral part of the *Investigations* curriculum. For a teacher using these units, assessment is an ongoing process. You observe students' discussions and explanations of their strategies on a daily basis and examine their work as it evolves. While students are busy recording and representing their work, working on projects, sharing with partners, and playing mathematical games, you have many opportunities to observe their mathematical thinking. What you learn through observation guides your decisions about how to proceed. In any of the units, you will repeatedly consider questions like these:

- Do students come up with their own strategies for solving problems, or do they expect others to tell them what to do? What do their strategies reveal about their mathematical understanding?

- Do students understand that there are different strategies for solving problems? Do they articulate their strategies and try to understand other students' strategies?

- How effectively do students use materials as tools to help with their mathematical work?

- Do students have effective ideas for keeping track of and recording their work? Does keeping track of and recording their work seem difficult for them?

You will need to develop a comfortable and efficient system for recording and keeping track of your observations. Some teachers keep a clipboard handy and jot notes on a class list or on adhesive labels that are later transferred to student files. Others keep loose-leaf notebooks with a page for each student and make weekly notes about what they have observed in class.

Assessment Tools in the Unit

With the activities in each unit, you will find questions to guide your thinking while observing the students at work. You will also find two built-in assessment tools: Teacher Checkpoints and embedded Assessment activities.

Teacher Checkpoints The designated Teacher Checkpoints in each unit offer a time to "check in" with individual students, watch them at work, and ask questions that illuminate how they are thinking.

At first it may be hard to know what to look for, hard to know what kinds of questions to ask. Students may be reluctant to talk; they may not be accustomed to having the teacher ask them about their work, or they may not know how to explain their thinking. Two important ingredients of this process are asking students open-ended questions about their work and showing genuine interest in how they are approaching the task. When students see that you are interested in their thinking and are counting on them to come up with their own ways of solving problems, they may surprise you with the depth of their understanding.

Teacher Checkpoints also give you the chance to pause in the teaching sequence and reflect on how your class is doing overall. Think about whether you need to adjust your pacing: Are most students fluent with strategies for solving a particular kind of problem? Are they just starting to formulate good strategies? Or are they still struggling with how to start? Depending on what you see as the students work, you may want to spend more time on similar problems, change some of the problems to use smaller numbers, move quickly to more challenging material, modify subsequent activities for some students, work on particular ideas with a small group, or pair students who have good strategies with those who are having more difficulty.

Embedded Assessment Activities Assessment activities embedded in each unit will help you examine specific pieces of student work, figure out what it means, and provide feedback. From the students' point of view, these assessment activities are no different from any others. Each is a learning experience in and of itself, as well as an opportunity for you to gather evidence about students' mathematical understanding.

The embedded assessment activities sometimes involve writing and reflecting; at other times, a discussion or brief interaction between student and teacher; and in still other instances, the creation and explanation of a product. In most cases, the assessments require that students *show* what they did, *write* or *talk* about it, or do both. Having to explain how they worked through a problem helps students be more focused and clear in their mathematical thinking. It also helps them realize that doing mathematics is a process that may involve tentative starts, revising one's approach, taking different paths, and working through ideas.

Teachers often find the hardest part of assessment to be interpreting their students' work. We provide guidelines to help with that interpretation. If you have used a process approach to teaching writing, the assessment in *Investigations* will seem familiar. For many of the assessment activities, a Teacher Note provides examples of student work and a commentary on what it indicates about student thinking.

Documentation of Student Growth

To form an overall picture of mathematical progress, it is important to document each student's work in journals, notebooks, or portfolios. The choice is largely a matter of personal preference; some teachers have students keep a notebook or folder for each unit, while others prefer one mathematics notebook, or a portfolio of selected work for the entire year. The final activity in each *Investigations* unit, called Choosing Student Work to Save, helps you and the students select representative samples for a record of their work.

This kind of regular documentation helps you synthesize information about each student as a mathematical learner. From different pieces of evidence, you can put together the big picture. This synthesis will be invaluable in thinking about where to go next with a particular child, deciding where more work is needed, or explaining to parents (or other teachers) how a child is doing.

If you use portfolios, you need to collect a good balance of work, yet avoid being swamped with an overwhelming amount of paper. Following are some tips for effective portfolios:

- Collect a representative sample of work, including some pieces that students themselves select for inclusion in the portfolio. There should be just a few pieces for each unit, showing different kinds of work—some assignments that involve writing, as well as some that do not.
- If students do not date their work, do so yourself so that you can reconstruct the order in which pieces were done.
- Include your reflections on the work. When you are looking back over the whole year, such comments are reminders of what seemed especially interesting about a particular piece; they can also be helpful to other teachers and to parents. Older students should be encouraged to write their own reflections about their work.

Assessment Overview

There are two places to turn for a preview of the assessment opportunities in each *Investigations* unit. The Assessment Resources column in the unit Overview Chart (pp. I-13–I-15) identifies the Teacher Checkpoints and Assessment activities embedded in each investigation, guidelines for observing the students that appear within classroom activities, and any Teacher Notes and Dialogue Boxes that explain what to look for and what types of student responses you might expect to see in your classroom. Additionally, the section About the Assessment in This Unit (p. I-18) gives you a detailed list of questions for each investigation, keyed to the mathematical emphases, to help you observe student growth.

Depending on your situation, you may want to provide additional assessment opportunities. Most of the investigations lend themselves to more frequent assessment, simply by having students do more writing and recording while they are working.

Turtle Paths

Content of This Unit   Students explore problems involving paths, lengths of paths, perimeter, and turns while engaging in computer activities using the computer program *Geo-Logo* as well as noncomputer activities. They create commands to describe paths, including measures of distance and turns in paths. Students define triangles and have the *Geo-Logo* turtle draw equilateral triangles. They solve missing measures problems by finding unspecified length and turn measures in figures. They solve problems involving specified perimeter lengths: drawing rectangles (and other paths) each having a total path length of 200 turtle steps and writing procedures to design a face on the computer.

Note: This unit makes substantial use of computers. It is important to read Managing the Computer Activities in This Unit (p. I-21) and go through the *Geo-Logo* Teacher Tutorial (p. 87) prior to teaching this unit.

Connection with Other Units   If you are doing the full-year *Investigations* curriculum in the suggested sequence for grade 3, this is the eighth of ten units. Your class will have already had some experience using the computer in the 2-D Geometry unit, *Flips, Turns, and Area*. They will have done some linear measurement using different standard and nonstandard units in the Measuring and Data unit, *From Paces to Feet*. Students may also have had experience with *Geo-Logo* in second grade units *Introduction to Mathematical Thinking at Grade 2* and *How Long? How Far?*

If your school is not using the full-year curriculum, this unit can also be used successfully at grade 4. Work with *Geo-Logo* is extended in the grade 4 2-D Geometry unit, *Sunken Ships and Grid Patterns*.

Investigations Curriculum ■ Suggested Grade 3 Sequence

Mathematical Thinking at Grade 3 (Introduction)

Things That Come in Groups (Multiplication and Division)

Flips, Turns, and Area (2-D Geometry)

From Paces to Feet (Measuring and Data)

Landmarks in the Hundreds (The Number System)

Up and Down the Number Line (Changes)

Combining and Comparing (Addition and Subtraction)

▶ *Turtle Paths* (2-D Geometry)

Fair Shares (Fractions)

Exploring Solids and Boxes (3-D Geometry)

Investigation 1 ▪ Paths and Lengths of Paths

Class Sessions	Activities	Pacing
Session 1 (p. 5) WALKING PATHS	Walking and Talking About Paths Giving Commands to Walk Different Paths Off-Computer Choices: Working with Paths Homework: Paths at Home Extension: Paths at School Extension: More Maze Path Games	minimum 1 hr
Session 2 (p. 14) COMMANDING THE TURTLE	Introducing Work Procedures On-Computer Activity: Get the Toys	minimum 1 hr
Sessions 3 and 4 (p. 20) MAZES AND MAPS	More About *Geo-Logo* Teacher Checkpoint: Thinking Mathematically About Paths Sharing Solutions to Get the Toys Homework: Many Possible Paths Extension: Exchanging Toys Extension: Visualizing Paths Extension: Measurement and Skip Counting	minimum 2 hr

Mathematical Emphasis

- Understanding paths as representations or records of movement

- Describing paths with mathematical language (for example, "closed," "corner")

- Finding several paths that meet a set of geometric constraints

- Using *Geo-Logo* commands to construct paths and describe their properties

- Applying mathematical processes such as addition, subtraction, estimation, and "undoing" to paths

Assessment Resources

Teacher Checkpoint: Thinking Mathematically About Paths (p. 23)

Thinking About Paths (Dialogue Box, p. 26)

Arithmetic and Undoing (Dialogue Box, p. 27)

Materials

Computers

Apple Macintosh disk, *Geo-Logo*

Large-screen monitor

Masking tape

Small counters or cubes

Colored pencils or crayons

Overhead projector and transparencies

Student Sheets 1–6

Teaching resource sheets

Family letter

Investigation 2 ▪ Turns in Paths

Class Sessions	Activities	Pacing
Sessions 1 and 2 (p. 31) TURNS	Turning Your Body On-Computer Activity: Feed the Turtle Off-Computer Choices: Estimating Turns Sharing Your Work Homework: As the World Turns Extension: Turtle Tells Extension: Magazine Turns Extension: Drawing Paths	minimum 2 hr
Session 3 (p. 41) TURNS, TURTLES, AND TRIANGLES	Describing Triangles Off-Computer Choice: Tricky Triangles Homework: Triangle Cat Extension: Triangle Drawings	minimum 1 hr
Session 4 (p. 45) EQUILATERAL TRIANGLES	Discussing Triangles On-Computer Activity: Triangles Assessment: Off-Computer Choice: Writing About Triangles Extension: Closed Paths Extension: Triangles and Almost Triangles	minimum 1 hr
Sessions 5 and 6 (p. 54) MISSING MEASURES	On-Computer Activity: Finding Missing Measures Off-Computer Choices: Completing Shapes Teacher Checkpoint: Missing Measures Whole-Class Discussion: Missing Measures Homework: More Missing Measures Extension: Finish the Figure Game Extension: Largest Rectangle	minimum 2 hr

◕ **Ten-Minute Math ▪ Lengths and Perimeters**

Mathematical Emphasis

- Using degrees to measure turns, especially full, half, and quarter turns; estimating turn measures in degrees
- Describing the properties of triangles; using the definition of triangles to decide whether or not figures fit the definition
- Identifying properties of equilateral triangles
- Using *Geo-Logo* commands to draw equilateral triangles by estimating turn measures and using trial-and-error strategies
- Applying mathematical processes, such as quantitative reasoning, mental arithmetic, and logic, to find missing measures of figures

Assessment Resources

Observing the Students (p. 37)

Turns and Angles (Teacher Note, p. 39)

Getting a Feel for Degrees and Turns (Dialogue Box, p. 40)

Assessment: Off-Computer Choice: Writing About Triangles (p. 48)

Strategies for Finding the Amount of Turn (Dialogue Box, p. 52)

Teacher Checkpoint: Missing Measures (p. 57)

Arithmetic in Geometry (Dialogue Box, p. 60)

Materials

Computers

Scissors

Nonpermanent marking pen

Overhead projector and transparencies

Student Sheets 7–17

Teaching resource sheets

Investigation 3 ▪ Paths with the Same Length

Class Sessions	Activities	Pacing
Sessions 1 and 2 (p. 65) THE 200 STEPS	Making Paths with 200 Steps Off-Computer Choices: 200 Steps and Rectangles On-Computer Activity: 200 Steps Discussing 200 Steps and Four 90° Turns Homework: Cutting and Combining Rectangles	minimum 2 hr
Sessions 3, 4, and 5 (p. 71) FACING PROBLEMS	Facing the Challenge On-Computer Activity: Geo-Face Assessment: Planning, Assembling, and Presenting Geo-Faces Choosing Student Work to Save Homework: Geo-Face Plans Extension: Switching Geo-Face Plans Extension: Perimeter Challenges	minimum 3 hr
Sessions 6 and 7 (Excursion*) (p. 80) DESIGNING A *GEO-LOGO* PROJECT	A *Geo-Logo* Project	minimum 2 hr

◑ **Ten-Minute Math** ▪ **Lengths and Perimeters**

*Excursions can be omitted without harming the integrity or continuity of the unit, but offer good mathematical work if you have time to include them.

Mathematical Emphasis

- Constructing geometric figures that satisfy given criteria, using analysis of geometric situations, arithmetic, and problem-solving strategies

- Comparing and connecting drawn paths to the *Geo-Logo* commands that created them in order to describe, analyze, and understand geometric figures

- Understanding that shapes can be moved in space without losing their geometric properties

- Estimating and measuring the perimeters of various objects

- Posing and solving original geometric problems

Assessment Resources

Assessment: Planning, Assembling, and Presenting Geo-Faces (p. 74)

Choosing Student Work to Save (p. 75)

Assessment: Planning, Assembling, and Presenting Geo-Faces (Teacher Note, p. 77)

Materials

Computers
Drinking straws
String
Rulers
Calculators
Square tiles
Printer
Student Sheets 18–24

Following are the basic materials needed for the activities in this unit. Many of the items can be purchased from the publisher, either individually or in the Teacher Resource Package and the Student Materials Kit for grade 3. Detailed information is available on the *Investigations* order form. To obtain this form, call toll-free 1-800-872-1100 and ask for a Dale Seymour customer service representative.

Computers—Macintosh II or above, with 4 MB of internal memory (RAM) and Apple System Software 7.0 or later. Maximum: 1 for every 2 students. Minimum: 1 for every 4–6 students. It is possible to modify the unit for fewer computers (see Managing the Computer Activities in This Unit, p. I-21).

Apple Macintosh disk, *Geo-Logo*™, for *Turtle Paths* (packaged with this book)

A large-screen monitor on one computer for whole-class viewing (recommended)

Masking tape

Small counters or cubes: 10–15 per student

Number cubes: 8–10

Rulers: 8–10

Calculators: 8–10

Square color tiles: 6 per student (optional)

Colored pencils or crayons for each group of students

Nonpermanent marking pen

Overhead projector

Drinking straws: 3–4 per pair of students

String: 30-inch length per pair of students

Printer (for Excursion)

Scissors: 1 per student

The following materials are provided at the end of this unit as blackline masters. A Student Activity Booklet containing all student sheets and teaching resources needed for individual work is available.

Family Letter (p. 132)

Student Sheets 1–24 (p. 133)

Teaching Resources:

   Geo-Logo User Sheet (p. 139)

   Dot Paper (p. 140)

   360 Degrees (p. 152)

   Turtle Turners (p. 153)

Practice Pages (p. 163)

Related Children's Literature

Burningham, John. *Mr. Gump's Outing*. New York: Henry Holt, 1970.

Jonas, Ann. *Round Trip*. New York: Greenwillow Books, 1983.

Much power in mathematics comes from connecting two big ideas—number and geometry. This unit engages third graders in several activities that help them link these ideas.

We use the notion of *paths* in many areas of thinking—from a path in the woods, to a path of reasoning, to a path of least resistance. In this geometry unit, a path is a representation or record of movement. Paths are lines, curves, corners, and so on.

We can think of typical shapes such as triangles or rectangles as paths with special attributes, or properties. For example, a square is a simple, closed path with four straight line segments of the same length and four corners made by the same-size turns.

This way of thinking has several advantages. It emphasizes *action*. Piaget showed that children do not learn about shapes by taking "mental photographs" of them. They learn about shapes by moving their fingers or eyes (possibly the infant's first experience of shape) along the contours of the shapes. Students learn about shapes through two basic kinds of movements: forward (or backward) movements that create line segments and turns that help to create angles. A great number of geometric objects and problems can be built using these two movements.

In this unit, students experience shapes as dynamic paths. They also use measures of the shapes to integrate their knowledge of number and shape. They draw shapes by giving commands to a computer "turtle." This type of drawing—compared with drawing by hand—requires students to translate what they *see* and understand *intuitively* and *implicitly* into what they can *analyze mathematically* and so understand *explicitly*.

As students explore these ideas, they are learning mathematics. As they find multiple paths that meet certain criteria (for example, "Find a path through this maze that is 40 steps long and has 3 corners"), they are also discovering that mathematical problems can have many solutions. As they construct a definition for the idea of "triangle," they are also constructing the idea that mathematics is *not* just "knowing" the definition.

When they find the missing measures of geometric figures, they are also developing their skills in mental arithmetic and coming to understand that mathematics is about connections between topics. When they design a computer program to draw a face, they are applying number and geometric knowledge, as well as seeing that mathematics is creative. When they explore these ideas on the computer, they are also discovering that the computer is not just something "to learn" but also a tool for mathematical exploration and thinking and for self-expression.

Mathematical Emphasis At the beginning of each investigation, the Mathematical Emphasis section tells you what is most important for students to learn about during that investigation. Many of these mathematical understandings and processes are difficult and complex. Students gradually learn more and more about each idea over many years of schooling. Individual students will begin and end the unit with different levels of knowledge and skill, but all will gain greater knowledge about two-dimensional space and shape and develop strategies for solving problems involving these ideas.

Throughout the *Investigations* curriculum, there are many opportunities for ongoing daily assessment as you observe, listen to, and interact with students at work. In this unit, you will find two Teacher Checkpoints:

>Investigation 1, Sessions 3–4:
>Thinking Mathematically About Paths (p. 23)

>Investigation 2, Sessions 5–6:
>Missing Measures (p. 57)

This unit also has two embedded assessment activities:

>Investigation 2, Session 4:
>Off-Computer Choice: Writing About Triangles (p. 48)

>Investigation 3, Sessions 3–5:
>Planning, Assembling, and Presenting Geo-Faces (p. 74)

In addition, you can use almost any activity in this unit to assess your students' needs and strengths. Listed below are questions to help you focus your observation in each investigation. You may want to keep track of your observations for each student to help you plan your curriculum and monitor students' growth. Suggestions for documenting student growth can be found in the section About Assessment (p. I-10).

Investigation 1: Paths and Lengths of Paths

■ How do students connect movement with paths? Given the description of a movement, can they draw the path? Given a path, can they describe the corresponding movement?

■ What language do students use to describe paths? Can they show the difference between closed and open paths? How do they describe the difference between a turn and a curved line?

■ How do students search for paths that satisfy a set of constraints after they have already found one? Can they use one such path to generate others?

■ Can students use *Geo-Logo* commands fluently? Do they use right and left turns consistently, even when the turtle is not facing straight up? Can they visualize a shape by looking at the *Geo-Logo* commands that generate it?

■ Can students combine several adjacent commands (such as fd 10, fd 20, fd 10) into one command (fd 40) appropriately? How do they generate commands to reverse a path?

Investigation 2: Turns in Paths

■ Can students describe a turn using the number of degrees in it (for example, 180° is a half turn)? What strategies do students use to figure out how many degrees are in, for example, a 3/4 turn? To figure out how big a turn of 100° is?

■ How do students describe triangles? How do they determine whether a given figure is or is not a triangle? Does changing the orientation of a triangle (for example, so that it does not have a horizontal base) affect their ability to recognize it?

■ How do students determine whether a triangle is equilateral when they look at a shape that has been drawn? When they look at *Geo-Logo* commands for drawing a shape?

■ Can students look at the results of a sequence of *Geo-Logo* commands and figure out how to change them to come closer to the figure they want to draw? Do they know whether to make a turn larger or smaller to achieve the desired figure?

■ What clues do students make use of in finding missing measures in figures? Do they notice and use parallel lines of the same length (as in opposite sides of a rectangle)? Do they add up several line segments to arrive at the length of a single line?

Investigation 3: Paths with the Same Length

■ How do students go about constructing geometric figures that have a specified number of sides? That have a specified perimeter? How do they integrate visual and numerical strategies?

■ How do students generate *Geo-Logo* commands to match a drawn figure? How do they decide on the number of steps for each side? Do they make use of comparisons between sides to decide on the number of steps for sides? (For example, this side is twice as long as that one.)

■ Can students see that two geometric figures are the same size and shape (congruent) even if they are in different locations and have different orientations? Can they recognize that two geometric figures are the same shape (for example, a square) even if they are different sizes?

■ How do students figure out the perimeter of a figure? Do they make use of their knowledge of the number system to add the lengths of the sides? Can they estimate perimeters by comparing a figure with sides of unknown lengths to one with sides of known lengths?

■ How do students tackle a complex problem such as creating a drawing that meets several different criteria? Can they integrate their solutions to separate parts of the problem, such as various shapes in a drawing that need positioning in particular ways?

In the *Investigations* curriculum, mathematical vocabulary is introduced naturally during the activities. We don't ask students to learn definitions of new terms; rather, they come to understand such words as *factor*, *area*, and *symmetry* by hearing them used frequently in discussion as they investigate new concepts. This approach is compatible with current theories of second-language acquisition, which emphasize the use of new vocabulary in meaningful contexts while students are actively involved with objects, pictures, and physical movement.

Listed below are some key words used in this unit that will not be new to most English speakers at this age level but may be unfamiliar to students with limited English proficiency. You will want to spend additional time working on these words with your students who are learning English. If your students are working with a second-language teacher, you might enlist your colleague's aid in familiarizing students with these words before and during this unit. In the classroom, look for opportunities for students to hear and use these words. Activities you can use to present the words are given in the appendix, Vocabulary Support for Second-Language Learners (p. 85).

closed, open In Session 1 of Investigation 1, students walk and describe masking-tape paths on the floor. One way they categorize paths is as *open* or *closed*.

corner Beginning with the first session, students describe paths and shapes. They talk about the *corner* of a path or shape and turning a corner to draw a path.

command In Session 1, students give *commands* to one another to walk a path. Throughout the unit, students write commands on the computer for the *Geo-Logo* Turtle to follow to draw paths or shapes.

This unit is dependent on having students use computers on an on-going basis. Ideally, students, working in pairs, will use computers daily for approximately 20 (or more) minutes. This means that you will need to plan carefully so all students have ample time to do the computer activities.

We have structured this unit as if you have five to eight computers in your classroom, enough so that half of your students, working in pairs, can use them at one time. At present, few classrooms have access to this many computers, but we have chosen to write the unit in this way for two reasons: First, having five to eight computers available all day in your classroom is most conducive to effective and efficient use of these resources. Second, writing the unit in this way provides a model for the way computers may be integrated in classrooms in the near future.

We have also provided advice below on how to modify the management of computer resources in two other common situations: where there are fewer than five computers available during math period and where there is a computer laboratory available outside the classroom.

Structuring the Unit to Match Computer Availability

Five to Eight Computers With five to eight computers, half the class, working in pairs, can use them at once. In each session, following a whole-class discussion, half the students do an On-Computer Activity and the other half do Off-Computer Choices. The two groups switch halfway though the work time. (Continuing to allow pairs of students to work at the computers throughout the school day will provide opportunities for students to complete all the computer activities plus time for them to redo some activities using different commands or strategies.)

Computer Laboratory If you have a computer laboratory, you may wish to involve the whole class in computer activities at the same time. Specific suggestions for modifying the sessions for a computer laboratory are found in the overview for each investigation. (See pp. 4, 29, and 63.)

Fewer than Five Computers If you have fewer than five computers available to use during the

math period, it is *mandatory* that students rotate using the computers throughout the school day, so every pair of students has completed the activities before the follow-up discussion. With this strategy, you will always introduce computer activities before the off-computer activities. Students can begin cycling through the computer activity as you work with the remainder of the class on the Off-Computer Choices. Specific suggestions for how to do this are included in the overview for each investigation.

Working at the Computer

Working in Pairs Students should work in pairs on the computers. Working in pairs not only maximizes computer resources but also encourages students to consult, monitor, and teach one another. Generally, more than two students at one computer find it difficult to share. (If you have an odd number of students, a threesome can be formed.) Since students frequently will not complete their computer work in one session, we suggest students stay with the same partner for the entire unit.

Saving Student Work Students will need to save their work on the computer activities. This can be done in two ways: (1) Students can use the same computer each time and save their work on the computer's internal drive, or (2) students can save their work on their own disk. If you can provide pairs of students with their own disks, computer management will be simpler, since each pair will be able to use any computer when it is available. If students' work is saved on a particular computer, they may have to wait until other students using that computer are finished with their work. Instructions for saving work are on p. 97 of the *Geo-Logo* Teacher Tutorial.

Demonstrating Computer Activities

Frequently you will need to use a computer with the whole class to demonstrate computer activities and to share results during whole-class discussions. It is helpful if a computer is connected to a large-screen monitor or projection device—a "large display." If you do not have a large display available, we suggest you gather groups of students as close as possible around the computer. Increasing the font size when entering commands will make them more visible for any demonstrations. To increase the font size, choose **All Large** under the

Font menu. (When you are finished demonstrating, remember to return the font to its regular size by choosing **All Small** under the **Font** menu.)

If your computer display is very small and it is difficult for your students to see the computer demonstrations, you may want to make a transparency of some of the student sheets with the computer screens to show students the commands you are entering.

Investigations

Paths and Lengths of Paths

What Happens

Session 1: Walking Paths Session 1 begins with two whole-class activities. Students walk on masking-tape paths on the floor, describing the movements they make. Students create paths for others by giving them movement commands (such as "forward 5 steps, left turn 90, forward 2 steps"). The session continues with the introduction of two or three Off-Computer Choices (Off-Computer Choice 3 can be introduced in Session 2). Students work in pairs to make paths on dot paper and to count steps in a maze to find different paths, such as one that is 14 steps in length and has 2 corners.

Session 2: Commanding the Turtle In Session 2, students are introduced to *Geo-Logo* and the first On-Computer Activity, Get the Toys. If they were not introduced to Off-Computer Choice 3, Maze Game, in Session 1, they are introduced to that activity during this session. Then students are divided into two groups—one working at the computers and the other working at their seats. The groups are switched halfway through the work period. Within each group, students work in pairs to find and describe paths. Students working at the computer give commands to the *Geo-Logo* turtle to fetch toys located in various positions on floor maps. Students working at their seats continue the Off-Computer Choices introduced in Sessions 1 and 2.

Sessions 3 and 4: Mazes and Maps Session 3 begins with a review of *Geo-Logo* and the On-Computer Activity, Get the Toys. It continues with an explanation of how to use *Geo-Logo* tools to "teach" and "run" procedures students have created. Session 3 continues with the class being divided into two groups. Working in pairs, one group continues to work on the On-Computer Activity, Get the Toys, and the other group continues with the Off-Computer Choices introduced in Sessions 1 and 2.

Session 4 begins after all students have had the opportunity to work at the computer. Students share their strategies and solutions to how they got the toys. Session 4 continues with the class being divided into two groups. Working in pairs, one group continues to work on the On-Computer Activity, Get the Toys, and the other continues to work on the Off-Computer Choices introduced in Sessions 1 and 2.

Mathematical Emphasis

- Understanding paths as representations or records of movement
- Describing paths with mathematical language (for example, "closed," "corner")
- Finding several paths that meet geometric constraints (for example, finding several ways to create a path that is 30 steps long with 2 corners)
- Using *Geo-Logo* commands to construct paths and describe their properties
- Applying mathematical processes such as addition, subtraction, estimation, and "undoing" to paths

What to Plan Ahead of Time

Materials

- Computers—Macintosh II or above, with 4 MB of internal memory (RAM) and Apple System Software 7.0 or later. Maximum: 1 for every 2 students. Minimum: 1 for every 4–6 students. It is possible to modify the unit for fewer computers. (See Managing the Computer Activities in This Unit, p. I-21).

- Apple Macintosh disk, *Geo-Logo*™, for *Turtle Paths* (Sessions 2–4)

- A large-screen monitor on one computer for whole-class viewing (recommended)

- Masking tape (Session 1)

- Small counters or cubes: 10–15 (Sessions 1–4)

- Number cubes: 8–10 (Session 1–4)

- Colored pencils or crayons for each group of students (Sessions 1–4)

- Overhead projector

Other Preparation

- Duplicate student sheets and teaching resources (located at the end of the unit) in the following quantities. If you have Student Activity Booklets, copy only the items marked with an asterisk, including any extra materials and transparencies needed.

For Sessions 1–4

Student Sheet 1, Maze Paths (p. 133): several copies per student, and 1 overhead transparency*

Student Sheet 2, Maze Paths Challenges (p. 134): 1 per student, and 1 overhead transparency*

Student Sheet 3, Paths at Home (p. 135): 1 per student (homework)

Geo-Logo User Sheet* (p. 139): 1 per computer

Dot Paper (p. 140): 1 sheet per student

Family letter* (p. 132): 1 per student. Remember to sign it before copying.

For Sessions 2–4

Student Sheet 4, Floors 1 and 2 (p. 136): several per student, and 1 overhead transparency*

Student Sheet 5, Floor 3 (p. 137): 1 per student

Student Sheet 6, Many Possible Paths (p. 138): 1 per student (homework)

- Use the disk for *Turtle Paths* to install *Geo-Logo* on each computer. (See pp. 128–129 in the *Geo-Logo* Teacher Tutorial.)

- Work through the following sections of the *Geo-Logo* Teacher Tutorial.

   Overview (p. 89)

   Getting Started with *Geo-Logo* (p. 91)

   Get the Toys

      How to Start an Activity (p. 93)

      How to Play Get the Toys on Floor 1 (p. 93)

      How to Save Your Game or Work (p. 97)

      How to Finish an Activity (p. 98)

Continued on next page

How to Continue with a
Saved Game (p. 99)

How to Play Get the Toys on
Floor 2 and Floor 3 (p. 99)

More About Get the Toys (p. 100)

More About *Geo-Logo* (p. 114)

■ Plan how to manage the computer activities.

If you have five to eight computers, have students work in pairs and follow the investigation structure as written.

If you are using a computer laboratory, in Sessions 1 and 2 have all students complete the Off-Computer Choices. At the end of Session 2, introduce the computer activity, Get the Toys, to the whole class. Dedicate Sessions 3 and 4 to having students do computer work in the lab.

If you have fewer than five computers, assign some students to the computer activity, Get the Toys, immediately after you introduce the activity in Session 2. Make and post a schedule to use during the last three days of the investigation. Assign about 15 to 20 minutes for each pair of students to use the computer(s) throughout the day.

■ Post the *Geo-Logo* User Sheet (p. 139) next to each computer. This sheet provides information about running the program and entering commands.

■ Prior to Session 1, use masking tape to lay out paths, as described in the activity Walking and Talking About Paths (p. 6).

■ If you plan to provide folders in which students will save their work for the entire unit, prepare these for distribution during Session 1.

Walking Paths

What Happens

Session 1 begins with two whole-class activities. Students walk on masking-tape paths on the floor, describing the movements they make. Students create paths for others by giving them movement commands (such as "forward 5 steps, left turn 90, forward 2 steps"). The session continues with the introduction of two or three Off-Computer Choices (Off-Computer Choice 3 can be introduced in Session 2). Students work in pairs to make paths on dot paper and to count steps in a maze to find different paths, such as one that is 14 steps in length and has 2 corners. Their work focuses on:

- walking paths and describing paths using mathematical ideas and language (for example, "closed," "straight," "corner," "angle," "turn")
- understanding paths as representations or records of movement
- describing movements that create paths and giving commands that create paths with certain shapes

The chart below shows how students work during this session.

Session 1	
Whole Class *40 min*	**Walking and Talking About Paths** **Giving Commands to Walk Different Paths** **Off-Computer Choices** (Introduction) ■ Dot Paths ■ Maze Paths ■ Maze Game
Students (working in pairs) *20 min*	**Off-Computer Choices** ■ Dot Paths ■ Maze Paths ■ Maze Game

Materials

- Student Sheet 1 (several copies per student)
- Transparency of Student Sheet 1
- Student Sheet 2 (1 per student)
- Transparency of Student Sheet 2
- Student Sheet 3 (1 per student, homework)
- Family letter (1 per student)
- Dot Paper (1 per student)
- Masking tape
- Small counters or cubes (10–15)
- Number cubes (8–10)
- Colored pencils or crayons (for each group of students)
- Overhead projector

Walking and Talking About Paths

Before the session, use masking tape to lay out paths large enough to walk on. Lay out the paths on as large a surface as possible (if there is no large area in the classroom, you might use the gym, the hallway, or an outside area). Below are some suggestions for the paths.

Have you ever walked a path? Describe it.

Examples might include paths in the snow or in sand, paths from their homes to school, and paths from the front door of the school to the classroom.

Tell students that a path, such as a beaten-down path through a woods, is a record of movement. Direct students' attention to the paths on the floor. Ask them to walk these paths. While they are doing so, encourage them to describe their movements (for example, they are going straight, they are turning at one place). (See the **Teacher Note**, Paths and Shapes, p. 11.)

Gather students around the paths on the floor. Stand at a beginning point on a path that consists of only line segments and corners. While they are walking the path, ask students what movements they must make (for example, forward some number of steps and turn right or left).

Ask how the masking-tape paths are the same as and different from one another. Have students explain (in terms of movement) straight paths (line segments), corners, straight and curved sections of paths, and closed paths (see the **Dialogue Boxes**, Paths with Turns, p. 12, and Is This Path Closed?, p. 13).

Have students stand on a path that has various properties (for example, closed, not closed, has line segments only).

Ask students to take their seats or return to the classroom.

Giving Commands to Walk Different Paths

We are going to write commands that will give directions for someone, or a robot, to walk a path. The commands tell the person or robot to move or turn. The move commands are *forward* or *back* [write fd and bk *on the board or a chart*]. The turn commands are *right* or *left* [write rt and lt *on the board or chart*]. Each command is followed by a number that tells how far to move, for fd and bk, or for how much to turn, for rt and lt.

❖ **Tip for the Linguistically Diverse Classroom** Quickly sketch a robot on the board and identify it before beginning the activity.

Ask a student to come to the front and be a robot, then say,

For example, if I want a robot to walk 5 steps forward, I would write fd 5.

Have the "robot" demonstrate the movement while you write the command on the board.

Now if I want our robot to turn and face the windows [*select something in the room that is 90° to the right of the robot*], **I would write** rt 90.

Again, have the "robot" demonstrate the turn while you write this command under the first command.

A 90-degree (90°) turn is like the right face we did in the Measuring and Data unit, *From Paces to Feet*, **and like commands given in the army or to bands marching in a parade.**

Write the command fd 5 on the board under the first two commands and ask the robot to follow it. (There should be three commands on the board: fd 5, rt 90, and fd 5.)

Ask another student to come to the board and draw the path the robot walked. For example: ⌐

Tell students they will play a game in which they will design a path for a robot, write commands for the path, and see if a robot can follow the commands. This path can be composed of no more than five straight line segments and can have only 90° turns or square corners (like the path just drawn on the board).

Identify a robot and an assistant and have them leave the room. Ask someone to draw a path on the board following the rules above. Work with the class to put together a series of commands that will teach the robot how to walk this path. Record the commands. Hide the path by covering it and have the robot and assistant return.

Read the commands and ask the robot to walk the path, with help from the assistant, who monitors which command is next. The class should note any problems with their commands and possible reasons for them. Ask the robot and assistant to draw on the chalkboard the path the robot actually walked while the students draw it at their seats. Then compare the drawings to the original path and discuss any discrepancies.

Repeat the activity with a new robot and assistant, possibly with some closed paths to check if the robot ends up where she or he started.

Off-Computer Choices: Working with Paths

Three Choices During Session 1 of this investigation, students choose one of three Off-Computer Choices available in the classroom. (During Sessions 2, 3, and 4, each student does an On-Computer Activity as well as an Off-Computer Choice.) Some students might try two Off-Computer Choices during the four sessions. Others might repeat an Off-Computer Choice. This format allows students to explore the same idea at different paces when they are not working at the computer.

Off-Computer Choice 1: Dot Paths

Students need dot paper and work as partners. First, each student writes a series of five commands such as fd 5, rt 90, fd 3, lt 90, bk 4. Then each student follows these commands on dot paper to make a path. Next, partners trade commands and make a dot-paper path following their partner's commands. Partners compare their pictures and discuss any discrepancies. They can vary the activity by walking through the commands on the floor.

Off-Computer Choice 2: Maze Paths

Give each student a copy of Student Sheet 1, Maze Paths, and show a transparency of the sheet on the overhead projector.

A girl starts on the Start dot. Each step she takes moves her the distance from one dot to the next. She walks seven steps into the maze. She is not allowed to double-back. Where could she be?

Have students tell where the girl is, justifying their answers. Discuss the various possibilities.

Responses might include the following: "7 dots straight ahead from the start." "Go left 5 dots, then go up 2 dots." Some students may have difficulty verbalizing and may point out various locations on the transparency. Encourage them also to try to describe the locations.

Repeat the situation with a different number of steps. The next time she comes to the maze, she walks 19 steps from the Start dot. Where could she be?

Again, have students tell where the girl could be, justifying their answers.

Give each student a copy of Student Sheet 2, Maze Paths Challenges, and show a transparency of the sheet. Explain that they are to use different colors to draw each path on the Maze Paths sheet (Student Sheet 1). They are not to draw too many paths on one sheet and may use more than one sheet of Maze Paths.

❖ **Tip for the Linguistically Diverse Classroom** Pair limited English proficient students with English proficient students to complete this worksheet. Encourage English proficient students to make the directions comprehensible to their partners by drawing quick sketches (for example, of a corner; or something open, closed, or crossing itself) as they read each instruction aloud.

Off-Computer Choice 3: Maze Game

Explain the directions to the Maze Game using the Maze Paths transparency (Student Sheet 1). Tell students they will need to choose and label a Finish dot on one copy of Student Sheet 1. They need counters and two number cubes. Students play with partners. Taking turns, a player rolls the two number cubes, adds the numbers, and moves his or her counter that many steps. The player cannot pass another player's counter. If the path is blocked, the player needs to look for a different path. The player who reaches the Finish dot first wins.

Classroom Management

Introduce two or three choices. (You may want to delay the introduction of Choice 3 until Session 2.) In whatever time remains, students begin working on one of the choices. (They may not have much time to work during Session 1.)

When students are doing the Off-Computer Choices, it is probably best to have them work at their seats, picking up the materials for the Off-Computer Choices at a central location, rather than have them work at stations (since the computers will be set up around the room). You can put student sheets in a folder labeled with the materials needed. For example:

Choice 1, Dot Paths: Dot paper

Choice 2, Maze Paths: Student Sheets 1 and 2

Choice 3, Maze Game: Student Sheet 1, to use as a game board; small counters or cubes; number cubes

 **Homework**

Send the family letter or the *Investigations* at Home booklet home today. Be sure to sign the letter before copying it.

Paths at Home   For homework, give each student a copy of Student Sheet 3, Paths at Home. Students choose two paths they often follow at home (for example, front door to refrigerator, bathroom to bedroom, and so on). For each path, students record the starting and ending points in the appropriate places on the student sheet, then write a series of robot commands to describe the path between them. If the path contains turns that are not 90°, students can describe them in words (for example, lt to face the windows).

Extensions

Paths at School   When students leave the classroom to go to different places in the building, have the line leaders describe the paths they are taking using the commands fd, rt 90, lt 90. Also have them tell how many corners they turned to get there.

More Maze Path Games   Suggest that students create some variations for the Maze Path Game. For example, they might draw a star on some of the dots on Student Sheet 1, and make a rule that if you land on a star dot, you take another turn.

Paths and Shapes

A path is a representation or record of movement. It is what you get if you move your pencil on a piece of paper without lifting it and without retracing a part of the path already drawn (however, a path can "cross itself").

In the activities in this unit, students become familiar with straight paths, corners, closed paths, and special paths such as polygons (triangles, squares, rectangles, and the like) and circles.

A *straight path* results from movement without turns. We often refer to such paths as "lines."

Lines and line segments are straight paths.

A *turn* is a rotation or change in direction and creates a corner in a path. A path with turns has corners, but not curves.

Paths that have curved sections

are created by continuous change in direction, or by going forward and turning at the same time.

Closed paths are those whose starting and ending points are the same.

A *shape* can refer to familiar figures, such as triangles and squares. More generally, it is the form or contour of anything. Used in this second way, every path has a shape. It is valuable to begin to refer to paths as figures or shapes so students do not mistakenly assume that a path is not a figure or shape.

Paths with Turns

This discussion took place as students were doing the activity Walking and Talking About Paths (p. 6).

The nature of turns in paths—which we'll return to again and again in the unit—often elicits interesting ideas right from the beginning.

Are there any paths you would consider similar in some way?

Maria: This one ⟨ and this one ⟨ are the same because they're almost like *S*'s.

What's different about them?

Khanh: The difference is that you had to turn on the first one.

Don't you turn on the curvy *S*?

Khanh: On the one with square corners you have to make bigger turns.

Samir: What?

Khanh: You turn a lot in one place.

Tamara: On the curvy one, you go right and left, real quick.

Explain that more.

Tamara: You turn right and right, then immediately left and left.

Are you just turning right or left?

Ryan: No, you're going forward and turning—at the same time.

What is a turn?

Elena: It's when you go around.

Yvonne: You change the direction you're heading.

What could you say about a path made without any turns?

Yvonne: It would have to be a straight path.

Is This Path Closed?

Determining whether a path or figure is closed is not always a simple matter. In one classroom, students had agreed on simple closed paths () and paths that were clearly not closed ( ). But they disagreed passionately on other paths ( ). The teacher thought that the intensity and engagement of students in the discussion was valuable and decided not to "provide the right answer" immediately. The teacher realized that students need time to grapple with intellectual problems and have their ideas and thoughts respected.

Sean: That's closed. There's a box right there.

Ricardo: It is not. Because a line sticks out.

Su-Mei: It's not because you don't end where you started.

Saloni: You can. Start here [*the left bottom of the square*] and go around and end here. Then come back to here.

Ricardo: That's retracing! You have to end where you start and not retrace.

Commanding the Turtle

Materials

- Choice Time materials from Session 1
- Student Sheet 4 (several per student)
- Transparency of Student Sheet 4
- *Geo-Logo* User Sheet (1 per computer)
- Computers with *Geo-Logo* installed
- Overhead projector

What Happens

In Session 2, students are introduced to *Geo-Logo* and the first On-Computer Activity, Get the Toys. If they were not introduced to Off-Computer Choice 3, Maze Game, in Session 1, they are introduced to that activity during this session. Then students are divided into two groups—one working at the computers and the other working at their seats. The groups are switched halfway through the work period. Within each group, students work in pairs to find and describe paths. Students working at the computer give commands to the Logo turtle to fetch toys located in various positions on floor maps. Students working at their seats continue the Off-Computer Choices introduced in Sessions 1 and 2. Their work focuses on:

- finding alternative solutions to path problems that pose specific geometric constraints
- using *Geo-Logo* commands to construct paths on a map to reach a given location
- applying mathematical processes such as addition, subtraction, estimation, and "undoing" to *Geo-Logo* commands to figure out the most efficient way to move on a map
- combining commands efficiently

The chart below shows how students work during this session.

Session 2		
Whole Class *30 min*	**Introducing Work Procedures** **On-Computer Activity: Get the Toys** (Introduction)	
Two Groups (working in pairs) *15 min*	Group A **On-Computer Activity** ■ Get the Toys	Group B **Off-Computer Choices** ■ Dot Paths ■ Maze Paths ■ Maze Game
Switch *15 min*	**Off-Computer Choices**	**On-Computer Activity**

Tell students that today they will be using computers for the first time in this unit. Depending on how you structure your class, explain how the class will operate during the remainder of the unit so everyone has an opportunity to do the computer work (see p. I-21, Managing the Computer Activities in This Unit).

For example, if you are following the chart above, you might state that:

■ the class will be divided into two groups.

■ each day the class work time will be divided into two equal blocks.

■ one group of students will work on the computer during the first block of time, and the other group will work at their seats; then the groups will switch.

■ each student will work with a partner.

■ students will work with the same partners for the entire unit.

■ students will use the same computer throughout the unit (or save their work on a disk so they can put it on another computer).

■ when working at computers, partners will take turns. First, one partner uses the mouse and the keyboard while the other reads and writes. They trade places after a period of time—for example, every 5 minutes.

Explain that at the beginning of each class session, you will introduce the On-Computer Activity and/or the Off-Computer Choices.

Note: It is critical that you have worked through the Overview (pp. 89–90), Getting Started with *Geo-Logo* (pp. 91–92), and Get the Toys (pp. 93–102) in the *Geo-Logo* Teacher Tutorial before working on the computer activities with students. Post copies of the *Geo-Logo* User Sheet next to each computer. Also, read the **Teacher Note**, Thinking Geometrically with *Geo-Logo* (see p. 19).

Today, we're going to instruct the turtle robot to walk a path in a computer game. In this game, called Get the Toys, there are three toys in a building. Each toy is on a separate floor, and it's the turtle's job to get the toys. The turtle robot has a small battery that allows it to carry out only a small number of commands before it runs out of energy. The battery gets recharged at the start of each floor. Your job is to write as few commands as possible so the turtle robot can get the toy on each of the three floors before it runs out of energy.

Gather students around the computer, preferably one with a large display, so they can all see the screen. Demonstrate how to:

- turn on the computer.
- open *Geo-Logo*.
- open Get the Toys.

Present the directions.

You must teach the turtle robot to get as many toys as possible. Remember, the battery runs out of energy if you enter too many commands. Your first command must be to go to the first floor.

Type floor 1 and press the **<return>** key.

Note: To make the commands easier for students to read while you are demonstrating computer activities, increase the font size by choosing **All Large** under the **Font** menu. (When you are finished demonstrating, remember to return the font to its regular size by choosing **All Small** under the **Font** menu.)

Introduce the four basic commands: fd, bk, rt 90, lt 90. State that the commands are posted on a *Geo-Logo* User Sheet beside each computer.

During the demonstration, if you think it will help your students to count the dots, distribute copies of Student Sheet 4, Floors 1 and 2, or put a transparency of floor 1 on the overhead.

Ask students to look at the screen. Point out the elevator and explain that in this activity the turtle's belly is right on top of the starting dot. Also explain that the dots are 10 turtle steps apart.

Ask students what commands to enter for the turtle to get the toy car. As you type them, announce every key you press. For example:

We want to go forward 30 turtle steps, so I'll type fd space 30, then press the <return> key.

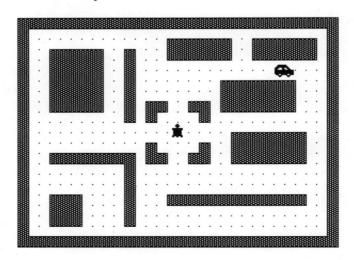

Enter commands students suggest, without evaluating them in any way.

Note: The main purpose of this demonstration is to show students how to use commands and tools in *Geo-Logo*. You don't want to "get the toy" in the demonstration during this session. After students have worked on the problem themselves, the class will discuss solutions in Session 3.

Sometimes when entering commands, pretend to forget to press **<return>** until the class reminds you to do so. Also, make a typing mistake and show how to use the **<delete>** key to erase a mistake.

Show how commands can be edited, such as changing fd 30 to fd 60.

I can change anything I already typed. I use the mouse or the arrow keys to move the cursor in the Command Center right after the 30, press the <delete> key two times to remove the 30, and type 60. When I press <return> I see the effect of that change—the turtle runs the new commands so the first line segment is 60 turtle steps long.

Show how commands can be combined, such as fd 40 fd 20, into a single command fd 60.

Also demonstrate how clicking on the **Erase One tool** at the top of the screen erases the last command.

Tell students that as they work at the computer, they are to record their commands on Student Sheet 4. Show how to do this by writing on a transparency or on the board the commands that have been entered on the computer.

Tell students that when they have found the toy, they are to write commands that will return the turtle to the elevator. It's OK to retrace their steps. When the turtle is back at the elevator, they are finished with floor 1 and should raise their hands. You will show them what to do next.

Tell students that when their time on the computers is nearly up, they will need to save their work so tomorrow they will be able to go on from where they left off. Demonstrate how to save their work by choosing **Save My Work** from the **File** menu and save the work on the computer (or disk), using their initials (or first or last name) and following the suggested procedure below. (Save your demonstration game for Session 3.)

You will want to save your work so you can begin where you left off the next time you use the computer. Your work will need a name so you can find it again. There are three things I want you to include in the name:

■ **the initials (or first or last names) of you and your partner(s)**

■ **a short one-word name for the activity**

■ **the date**

For example, if Cesar and Jeremy are working together, they could name their work, "CC + JW toys 3/22."

Classroom Management

Assign half the students to work at their seats on the Off-Computer Choices from Session 1 and assign the other half to work at computers on the On-Computer Activity, Get the Toys. (If students were not introduced to Off-Computer Choice 3, Maze Game, in Session 1, introduce it now.) Students work with partners. Remind them of the assignments.

Students working off the computer have three choices:

> Choice 1: Dot Paths
>
> Choice 2: Maze Paths
>
> Choice 3: Maze Game

If students working at their seats started an Off-Computer Choice yesterday and didn't finish it, they should complete it before selecting another choice.

Students working on the computer need to turn on the computer and open Get the Toys. The partners take turns at the computer entering the commands or writing them on Student Sheet 4; they switch about every five minutes. Encourage students to refer to the *Geo-Logo* User Sheet posted beside each computer.

Switch groups halfway through the work time.

While Students Are Working In Session 2, you will probably need to spend the major portion of your time helping students get started and enter commands on the computer. When they ask for assistance, answer their questions and refer them to the *Geo-Logo* User Sheet posted next to the computer.

If a pair of students completes floor 1, getting the toy car and returning to the elevator, show the pair how to teach the turtle its solution by choosing the **Teach tool** . Then demonstrate how to run the solution for floor 1 and go on to floor 2. (You will demonstrate the **Teach tool** to the whole class at the beginning of Session 3.)

Saving Work When it is near the time to switch groups, show students who are working at the computers how to save their work, either on a disk or on the computer's internal drive. (You may wish to gather them around one computer, demonstrating with one pair's work, then have each pair go back and save its own work.)

Also teach the second group of students how to save the work at the end of this session.

Thinking Geometrically with Geo-Logo

What benefit is there for students in using the computer program *Geo-Logo*? In what ways do explorations with *Geo-Logo* help students become more competent in geometric thinking? *Geo-Logo* is a special geometry-oriented version of Logo, a well-known computer language used internationally for teaching mathematics.

One major difference between *Geo-Logo* activities and pencil-and-paper geometry activities is that, in *Geo-Logo*, the emphasis is on constructing geometric paths, figures, and designs rather than on recognizing and naming them, as often is the case in standard curricula. *Geo-Logo* provides an environment in which students can easily create and modify geometric figures. They are engaged in constructing geometric figures using mathematical language.

A second major difference concerns the ways in which students verbalize the actions they perform. *Geo-Logo* requires them to communicate with the turtle in a geometrically oriented language. Thus, in the process of making squares, students see that the sides must be equal lengths because all four commands that draw sides have the same length.

The *Geo-Logo* language requires students to be precise, differentiating in writing between, for example, an 89-degree angle and a 90-degree one. This precision, which differs significantly from the much looser precision of drawings, also makes it possible to compare students' drawings of a square. While two hand-drawn squares can look similar even if the two students have different conceptions of a square, two *Geo-Logo* programs will highlight even small differences between students' procedures for drawing squares.

Geo-Logo programs can also be run again—and are guaranteed to produce the same drawing each time (assuming the turtle starts in the same place). Students can actually study what their programs do by running them one step at a time and fixing any command(s) they need to change.

Finally, computers in general, and *Geo-Logo* in particular, are motivating to students. The artistic aspect of some of the activities (for example, the face activity in Investigation 3) can provide students a different approach to geometric concepts.

Mazes and Maps

Materials

- Choice Time materials from Sessions 1 and 2
- Student Sheet 5 (1 per student)
- Computers
- Student Sheet 6 (1 per student, homework)
- Student Sheet 1 (1 per student, homework)

What Happens

Session 3 begins with a review of *Geo-Logo* and the On-Computer Activity, Get the Toys. It continues with an explanation of how to use *Geo-Logo* tools to "teach" and "run" procedures students have created. Session 3 continues with the class being divided into two groups. Working in pairs, one group continues to work on the On-Computer Activity, Get the Toys, and the other group continues their work on the Off-Computer Choices introduced in Sessions 1 and 2.

Session 4 begins after all students have had the opportunity to work at the computer. Students share their strategies and solutions to how they got the toys. Session 4 continues with the class being divided into two groups. Working in pairs, one group continues to work on the On-Computer Activity, Get the Toys, and the other group continues to work on the Off-Computer Choices introduced in Sessions 1 and 2. Their work focuses on:

- finding alternative solutions to paths problems that pose specific geometric constraints
- using *Geo-Logo* commands to construct paths on a map to reach a given location
- applying mathematical processes such as addition, subtraction, estimation, and "undoing" to *Geo-Logo* commands to figure out the most efficient way to move on a map
- combining commands efficiently

The following charts show how students work during these sessions.

Session 3		
Whole Class *20 min*	**More About *Geo-Logo***	
Two Groups (working in pairs) *20 min*	Group A **On-Computer Activity** ■ Get the Toys **Teacher Checkpoint: Thinking Mathematically About Paths**	Group B **Off-Computer Choices** ■ Dot Paths ■ Maze Paths ■ Maze Game
Switch *20 min*	**Off-Computer Choices**	**On-Computer Activity**

Session 4		
Whole Class *20 min*	**Sharing Solutions to Get the Toys**	
Two Groups (working in pairs) *20 min*	Group A **On-Computer Activity** ■ Get the Toys **Teacher Checkpoint: Thinking Mathematically About Paths**	Group B **Off-Computer Choices** ■ Dot Paths ■ Maze Paths ■ Maze Game
Switch *20 min*	**Off-Computer Choices**	**On-Computer Activity**

More About *Geo-Logo*

Gather students around the demonstration computer. Tell them that yesterday they had their first experience playing Get the Toys. It doesn't matter whether they finished floor 1. They will have more time to work on the activity during the next two days.

Let's review what we learned yesterday by doing floor 1 together.

Show students how to open the work they saved by opening your saved demonstration game. (Make sure the font size is **All Large**.) Have students suggest commands to finish getting the toy on floor 1. (If the procedure contains many extra or extraneous commands, you may wish to use

the **Erase All tool** 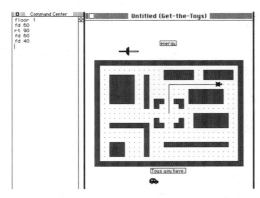 and begin again. You want the turtle commands to get the toy to be straightforward.)

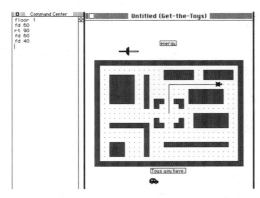

When the turtle has gotten the toy, tell students they now must return to the elevator and that it's OK to retrace their steps. Also tell them you're going to make it more difficult to return to the elevator in this demonstration than it will be when they are working at the computers. You are going to close the Drawing window so they can't see the turtle while they are giving you the commands to return the turtle to the elevator. (Select **Hide Drawing** under the **Windows** menu.)

Note: The reason for closing the Drawing window for the demonstration is to help students focus on the commands that got them to the toy rather than on the drawing. Most students will leave the Drawing window open while they work on the computer.

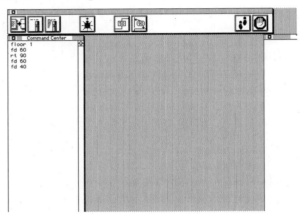

Now we can't see the turtle. Can you still tell me commands that will get the turtle back to the elevator? What would your strategy be?

Enter commands students suggest. Then choose **Show Drawing** on the **Windows** menu and check their solution.

When you have completed floor 1, demonstrate (review for some students) how to make these commands a procedure by choosing the **Teach tool** . Then demonstrate how to run the procedure by entering the name for the procedure in the Command Center. Emphasize that students should use the **Teach tool** only when they are finished with each floor.

Show students how to use the **Erase All tool** to clear the Command Center and the Drawing window before entering the command floor 2 to get the new floor plan.

The sequence of commands you have just demonstrated—**Teach, Run, Erase All**—is the general sequence students should follow after saving a computer problem in order to go to the next one.

Classroom Management

Divide the class into two groups. Have one group continue their work on the On-Computer Activity, Get the Toys. Have the other group continue working on the Off-Computer Choices. Put out Student Sheet 5, Floor 3, for students to use to record their computer commands for this floor. Switch groups halfway through the work time. Observe those students who are using the computer, keeping in mind the concepts and questions in the Teacher Checkpoint, Thinking Mathematically About Paths.

If students retrieved the toys on floors 1, 2, and 3, tell them there is a secret floor, floor 9. To get to it, they type floor 9 (even though the message is to type floor 1). Students can also solve Get the Toys again, trying to find on each floor the path to the toy that uses the fewest commands.

Activity

Teacher Checkpoint

Thinking Mathematically About Paths

As students are cycling through the On-Computer Activity, observe the strategies they use.

■ How are they determining how far to move the turtle?

■ Do they use the commands they wrote to get the toy to figure out what commands they should use on the return trip? How?

■ Are they combining commands appropriately (for example, combining fd 40 fd 20 into a single command, fd 60)?

■ Can they use right and left consistently, even when the turtle is not facing straight up?

See the **Dialogue Boxes,** Thinking About Paths (p. 26) and Arithmetic and Undoing (p. 27), for additional strategies, skills, and concepts to observe.

Interact with students, asking questions such as these:

How are you deciding what commands to enter? What command will you use next? Why?

If a pair seems to be having difficulty, you might suggest they work with another pair. Also, you might provide prompts and hints such as these:

What kind of path does fd 30 make?

What happens to the path when the turtle turns?

How did you make a corner in that path?

Which way will the turtle be facing if you type lt 90?

How can you get the turtle to face the toy?

For the Off-Computer Choices, observe whether students understand the rules and the properties of paths (for example, closed). Are they checking all alternative paths? Can they change the Maze Path game or make up their own game to add interest? If students need more Off-Computer Choices, have them work on the extensions on p. 25.

Activity

Sharing Solutions to Get the Toys

In Session 4, after all students have had an opportunity to work on Get the Toys at the computers, ask one pair to share their strategies and solution for floor 2 or floor 3 with the class. Have the pair describe the path from the elevator to the toy and how they returned to the elevator.

How many corners and straight sections does this path have from the elevator to the toy?

What commands did you use to make the straight sections, or line segments?

What commands did you use to make the corners?

Is the path closed?

Could you have taken a different route?

How long is each path?

How many commands did you give to the turtle to create each path?

Which paths are the best for Get the Toys—the shortest paths or the ones that use the fewest commands? Why?

Does the shortest path always use the least energy? (See the **Dialogue Box**, Thinking About Paths, p. 26.)

Before you try the challenge again, let's talk about good strategies. How can you use the fewest number of commands to make the battery last as long as possible?

(For example, combine `fd 10 fd 20` into `fd 30`; change `lt 90 rt 90 rt 90` to one command `rt 90`; search for the path that needs the fewest commands, even if it's not the shortest.) It may help students understand the concept if they enter the command `fd 1` ten or twenty times, then change it to `fd 20`. Ask them how each type of command affects the battery level. See the **Dialogue Box**, Thinking About Paths (p. 26).

Sessions 3 and 4 Follow-Up

Many Possible Paths   Send home Student Sheet 6, Many Possible Paths, along with a blank copy of Student Sheet 1, Maze Paths. Students draw several paths that fit each of two descriptions, drawing each path in a different color. Their paths may start at any dot on the page. After drawing three different paths for each description, they find and count (but do not draw) other paths that fit the same descriptions.

Homework

Exchanging Toys   Have groups exchange their written procedures for Get the Toys (Student Sheets 4 and 5). Without using the computer, each pair analyzes the other's procedure by (a) drawing the path the other pair made to get the toy, (b) determining the length of the path to the toy, and (c) figuring out whether they could make a path that would need fewer commands.

Extensions

Visualizing Paths   Tell students the following:

Close your eyes and visualize the answers to these questions: Could you make a closed path with three line segments? With two? Can you make a closed path with three 90° corners? A closed path with two 90° corners?

Measurement and Skip Counting   Students count by 10's to determine the lengths of the parts of the paths on their Get the Toys maps and by 5's to determine the lengths on their maze paths.

Thinking About Paths

Thinking aloud about the "best" path can help students clarify their ideas. Here, the teacher helps two students, working as partners, think about their approaches in the computer activity Get the Toys.

Which way would be the best route on floor 3?

Maya: Over the top.

Why?

Maya: It's the easiest.

Dylan: There's another way. [*He shows the bottom route, which actually uses fewer commands.*]

Maya: Yes, that uses fewer commands, but it takes up more space.

What wears down the battery?

Dylan: Fewer commands. No, no, more commands.

So which path are you going to take?

Students: This one. [*They point to the top route.*]

Why? [*She purposely begins repeating her questions.*]

Maya: It's the best. The easiest.

And the other route?

Maya: Uses fewer commands.

Dylan: But more commands run down the battery.

So which path are you going to take?

Students: This one. [*They point to the optimal bottom route!*] That's the one we're going to use. [*They show no signs of recognizing that they repeated virtually the same conversation but reversed their conclusions.*]

Maya: You want to take the shortest one.

Is it always true that a shortest path takes the least energy?

Dylan: Yes, even for a person walking.

Let me show you two paths and see which you think will take the most energy from the turtle.

Dylan: The one on the right, because there's lots of commands in it—here, here, and here—that are not in the other one. This one is a longer path, but it has fewer commands. The other one added turn commands.

What commands exactly were added?

Maya: A `rt 90` here, then a `fd` to make this straight line segment, then a `lt 90` here.

Arithmetic and Undoing

Here are excerpts from conversations at the computer where students were trying to use as few commands as possible to return the turtle to the elevator in Get the Toys.

How are you going to do this?

Midori: You should go 50. Because 30 + 10 + 10 is equal to 50.

Jamal: Go backward 50.

Why?

Midori: It's less commands to type one bk 50 than three backwards, 30, 10, and 10.

Elena: Let's try it again. We can make it this time.

Maya: How?

Elena: We'll use less commands. See? Here we'll go bk 50. Because, look, we put fd 30 in, but then we had to put in a fd 10 fd 10, so it's got to be fd 50.

[*A little later*]

So first you had lt 90 rt 90 rt 90. What did you change?

Dylan: We just used one rt 90. Otherwise, the turtle's just dancing!

How did that help you?

Dylan: It uses up the battery just once, not three times!

[*A little later*]

Maria: We have to get it back to the elevator. So do the same commands, but backwards. We have to go 80.

Rashad: No. 100. Because, see, we didn't get to the toy with fd 80. We didn't get there until we typed fd 10 and then fd 10 again! So, we have to go back 100.

Turns in Paths

What Happens

Sessions 1 and 2: Turns In Session 1, students turn their bodies and discuss ways to measure turns other than 90°. They learn about the Turtle Turner (protractor) to help them estimate and measure turtle turns (in multiples of 30°). Students are also introduced to an On-Computer Activity, Feed the Turtle, and three or four Off-Computer Choices (Choice 4 is optional).

In Session 2, students work in two groups—one group works on the On-Computer Activity and the other works at their seats on the Off-Computer Choices. Groups switch halfway through the work period. Within each group, students work in pairs. During the last 20 minutes of Session 2, students share their work with the class.

Session 3: Turns, Turtles, and Triangles In Session 3, students define triangles and use their definitions to determine whether various figures are triangles. They are introduced to one Off-Computer Choice, Tricky Triangles, which all students complete. Then students working at their seats may complete the Off-Computer Choices introduced in Session 1. Students working at the computer finish the On-Computer Activity, Feed the Turtle.

Session 4: Equilateral Triangles In Session 4, students discuss Tricky Triangles, applying their definitions of triangles. Students are introduced to an On-Computer Activity, Triangles. They identify equilateral triangles and write *Geo-Logo* procedures for drawing them. As an assessment, they write about triangles.

Sessions 5 and 6: Missing Measures In Session 5, students find missing lengths and turns needed to complete partially drawn figures. Students are introduced to an On-Computer Activity, Finding Missing Measures, and an Off-Computer Choice, Missing Lengths and Turns. All students begin work on the Off-Computer Choice, Missing

Lengths and Turns. Students work in pairs. As pairs complete their plans for one or two figures, they enter the commands for those figures on the computer.

In Session 6, students continue their work on the Off-Computer Choice, Missing Lengths and Turns, and on the On-Computer Activity, Finding Missing Measures. They are introduced to the Off-Computer Choice, Help Make Toys, and work on it. Session 6 ends with a whole-class discussion about missing measures.

Mathematical Emphasis

- Using degrees to measure turns; understanding that there are 360° in a full turn, 180° in a half turn, and 90° in a quarter turn; estimating turn measures in terms of degrees
- Describing the properties of triangles: closed figures with three straight sides and three corners; using the definition of triangles to decide whether or not figures fit the definition
- Identifying properties of equilateral triangles: all three sides are equal in length and all three turns are equal in measure
- Using *Geo-Logo* commands to draw equilateral triangles by estimating turn measures and using trial-and-error strategies
- Applying mathematical processes, such as quantitative reasoning, mental arithmetic, and logic, to find missing measures of figures

What to Plan Ahead of Time

Materials

- Computers (Sessions 1–6)
- Scissors: 1 per student (Sessions 1–2, optional)
- Nonpermanent marking pen
- Overhead projector

Other Preparation

- Duplicate student sheets and teaching resources as follows. If you have Student Activity Booklets, copy only the transparencies marked with an asterisk.

For Sessions 1–2

Student Sheet 7, Turning the Turtle (p. 141): 1 per student

Student Sheet 8, Turn Commands (p. 142): several per student

Student Sheet 9, As the World Turns (p. 143): 1 per student

Student Sheet 10, Feed the Turtle Commands (p. 144): 1 per student (optional)

360 Degrees* (p. 152): 1 overhead transparency

Turtle Turners* (p. 153): 1 transparency for every 4 students (cut each transparency into 4 Turtle Turners)

For Session 3

Student Sheet 11, Tricky Triangles (p. 145): 1 per student

Student Sheet 12, Which Are Triangles? (p. 146): 1 per student

Student Sheet 13, Triangle Cat (p. 147): 1 per student (homework)

For Session 4

Student Sheet 11*, Tricky Triangles (p. 145): 1 overhead transparency

Student Sheet 14, Writing About Triangles (p. 148): 1 per student

For Sessions 5–6

Student Sheet 15, Missing Lengths and Turns (p. 149): 1 per student

Student Sheet 16, Help Make Toys (p. 150): 1 per student

Student Sheet 17, More Missing Measures (p. 151): 1 per student (homework)

- Work through the following sections of the *Geo-Logo* Teacher Tutorial.

    Feed the Turtle

    How to Choose a New Activity (p. 103)

    How to Play Feed the Turtle (p. 103)

    Triangles

    How to Make Triangles (p. 105)

    More About Triangles (p. 106)

    Missing Measures

    How to Find Missing Measures (p. 108)

    More About Missing Measures (p. 108)

- Plan how to manage the computer activities.

    If you have five to eight computers, have students work in pairs and follow the Investigation structure as written.

    If you are using a computer laboratory, students will need to use the lab for four or

Continued on next page

five days for this investigation and could do part of the investigation in the classroom: You could do Session 1 in your classroom, introducing turns and the Turtle Turner, and have students work on the Off-Computer Choices. Beginning with Session 2, students will need one or two days in the lab to do the On-Computer Activity, Feed the Turtle. You could have students do the whole-group activities from Sessions 3 and 4— Describing Triangles, including Tricky Triangles (and the discussion of Student Sheets 11 and 12), Discussing Triangles, and the assessment, Writing About Triangles— in your classroom. In Sessions 5 and 6, you may wish students to do less extensive planning and checking off-computer, giving them more time to refine their initial plans while working with *Geo-Logo* in the computer lab.

If you have fewer than five computers, you could cycle students through the computer activities as described in Managing the Computer Activities in This Unit (p. I-21).

For Sessions 1 and 2, start some of the students on the computer immediately after demonstrating Feed the Turtle, then introduce the Off-Computer Choices to the rest of the class. Pairs of students continue cycling through the computer activity throughout the day (or the next two days), until every student has completed Feed the Turtle. Conduct whole-group parts of Sessions 3 and 4 with the whole class, then have pairs of students cycle through the computer activity, Triangles, while the others work on Off-Computer Choices. For Sessions 5 and 6, have pairs of students work together. They should solve the missing lengths problems on paper, then cycle through on the computer to try their solutions.

Turns

What Happens

Materials

- Transparency of 360 Degrees
- Turtle Turner transparencies (1 per student)
- Student Sheet 7 (1 per student)
- Student Sheet 8 (several per student)
- Student Sheet 9 (1 per student)
- Student Sheet 10 (1 per student, optional)
- Scissors (1 per student, optional)
- Nonpermanent marking pen
- Overhead projector
- Computers

In Session 1, students turn their bodies and discuss ways to measure turns other than 90°. They learn about the Turtle Turner (protractor) to help them estimate and measure turtle turns (in multiples of 30°). Students are also introduced to an On-Computer Activity, Feed the Turtle, and three or four Off-Computer Choices (Choice 4 is optional).

In Session 2, students work in two groups—one group works on the On-Computer Activity and the other works at their seats on the Off-Computer Choices. Groups switch halfway through the work period. Within each group, students work in pairs. During the last 20 minutes of Session 2, students share their work with the class. Their work focuses on:

- understanding turns as a change in orientation or heading
- exploring what happens when turns are repeated (such as when you perform four 90° turns you end up facing in the same direction as when you started)
- becoming familiar with a common measurement for turns—degrees. Understanding that there are 360° in one full turn, 180° in a half turn, and 90° in a quarter turn.
- estimating turn measures

The following charts show how students work during these sessions.

Session 1		
Whole Class *40 min*	**Turning Your Body** **On-Computer Activity: Feed the Turtle** (Introduction) **Off-Computer Choices:** (Introduction) ■ Turning the Turtle ■ As the World Turns ■ Turn Commands ■ Feed the Turtle Commands	
Two Groups (working in pairs) *10 min*	Group A **On-Computer Activity** ■ Feed the Turtle	Group B **Off-Computer Choices** ■ Turning the Turtle ■ Turn Commands ■ As the World Turns ■ Feed the Turtle Commands
Switch *10 min*	**Off-Computer Choices**	**On-Computer Activity**

Session 2		
Two Groups (working in pairs) *20 min*	**Group A** **On-Computer Activity** ■ Feed the Turtle	**Group B** **Off-Computer Choices** ■ Turning the Turtle ■ Turn Commands ■ As the World Turns ■ Feed the Turtle Commands
Switch *20 min*	**Off-Computer Choices**	**On-Computer Activity**
Whole Class *20 min*	**Sharing Your Work**	

 Ten-Minute Math: Lengths and Perimeters Ten-Minute Math activities are done outside of math time during any spare ten minutes you have during the day, perhaps right before lunch or at the end of the day. A few times during the next few days, do the Ten-Minute Math activity Lengths and Perimeters (it is an off-computer problem using the *Geo-Logo* repeat command).

Note: The repeat command is not formally presented in this unit, although it is available for students to use on the computer. If your students have not used the repeat command, you will need to introduce it the first time you do this activity. For hints, see the complete directions for Ten-Minute Math (p. 82).

Choose a distance you want the turtle to go, let's say 35. Ask students to work in pairs for 2 to 3 minutes and write down as many repeat commands as they can think of that would move the turtle that distance. They use the format repeat __ [fd __]. They can use calculators to test their ideas.

Make a list on the board or overhead of the different responses.

Ask students to prove their responses work:

How do you know repeat 5 [fd 7] would be 35 turtle steps?

Discuss with students whether they have all the possibilities. For example:

Do we have all the repeat commands that work? How do you know? Could we do anything with repeat 3? Could we do anything using fd 8?

For variations on this activity, see p. 82.

Turning Your Body

What is a turn?

Students may suggest that a turn is a change in direction or heading.

Suppose a robot was facing the front of the room, and I wanted it to turn and face the back of the room. What commands could I write?

Students might state rt 90 followed by another rt 90 (or lt 90 lt 90). A couple of students might suggest rt 180. If students are having difficulty understanding the idea, you might demonstrate and say,

I begin facing the front, and I make a lt 90 and another lt 90 [*turning and pausing between turns*]. I am now facing the back of the room. But what if I make a fast nonstop turn [*demonstrating*], can I write that with one command?

Have students stand at their seats, facing the front of the room, and figure out the answers to the following questions by turning their bodies. Ask students to share their reasoning following each question.

Do a rt 90 and another rt 90. What is one more command that will have us face the front of the room again?

How many rt 90 turns do I have to make to turn all the way around?

If I make three lt 90, how can I write that in one command?

(For the last question, accept both lt 270 and rt 90. Have students discuss why they end up in the same position.)

Ask students to visualize and point to where they would be after 1, 2, 3, or 4 rt 90 or lt 90 turns. Then have them perform the turns to check. Here, and in the following activities, occasionally have students close their eyes when they turn.

Show the 360 Degrees transparency (p. 152). Tell students that the turtle makes 360 very small turns to turn all the way around. The amount of each turn is called a degree—a right turn 90 is a turn of 90 degrees.

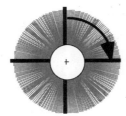

In the next computer activity, Feed the Turtle, students will be able to command the *Geo-Logo* turtle to make turns other than 90 degrees. This will allow them to draw shapes with corners that are not square corners.

Turns Less than 90° Introduce turns that are smaller than right turns by stating the following:

What if I wanted to make a turn that was smaller than a 90° turn? For example, how much would I turn if I wanted to face the corner of the room instead of the windows?

Accept students' suggestions. Most should agree that the turn will be a number smaller than 90.

Ask students to stand and pretend there is a huge clock on the floor, with the 12 at the front of the room and the 6 at the back. Have students be the hands in the center of the clock. Have them extend both arms to the front (12:00), rotate one arm to a 3:00 position, and then turn their bodies (keeping arms stationary) to show a rt 90.

❖ **Tip for the Linguistically Diverse Classroom** To ensure that limited English proficient students understand they are pretending to be a clock on the floor, indicate the classroom clock.

Ask students to again face front with their arms extended at 12:00, move one arm to 1:00, then turn their bodies to face one o'clock.

If I wanted to write a command to turn this much, what should I write?

Accept ideas from students, asking them to state their reasons. It may be helpful for students to turn their bodies through several hours of the clock, identifying appropriate turtle commands for larger angles before figuring out the command for moving from 12:00 to 1:00.

> "If I turn from 12:00 to 6:00, the command is rt 180."

> "If I turn from 6:00 to 4:00, the command is lt 60."

Introduce the Turtle Turner Show a Turtle Turner on the overhead projector. Students describe it, paying particular attention to the direction the turtle is facing, the distinction between lt and rt, and the symmetry between the left and right sides.

If any of your students have used the grade 2 *Investigations* unit *How Far? How Long?,* they will have seen a Turtle Turner that is similar but not identical. The grade 2 turner is different in two ways. There are fewer possible turns marked on each turner, and the possible turns are marked 1, 2, 3, 4, and 5 instead of 30, 60, 90, 120, 150. These simplifications are appropriate for grade 2, where students are focusing on the basic ideas that turns come in different sizes and that it is possible to compare turns to see which is larger. In grade 3, students move on to assigning degree measures to turns, based on dividing a circle into 360°. You should be aware of the similarities and differences between the two Turtle Turners so that you can answer any questions that may arise from students.

Remove the Turtle Turner, and on the overhead draw a line with a small turtle at the end. Demonstrate how to place the Turtle Turner in the same direction as the turtle and with the center of the Turtle Turner directly on the end of the line to determine the amount of turn. Ask students to sweep out the angle through which the turtle turns to illustrate 30°, 90°, and 120° turns. (See the **Teacher Note**, Turns and Angles, p. 39).

Note: In this unit, we emphasize turns that are multiples of 30°—30°, 60°, 90°, 120°, and so on.

Activity

On-Computer Activity: Feed the Turtle

Gather students around a computer to demonstrate Feed the Turtle. Explain to them how to open the activity as you open it. (See the *Geo-Logo* Teacher Tutorial, p. 103.)

The turtle is looking for food. You are to help the turtle get to the food before it runs out of energy. When it reaches food, it eats and gets more energy. Make sure you plan the order in which the turtle eats the food so it doesn't run out of energy. In this activity, for all the fd's or bk's, use multiples of 10—10, 20, 30, 40, 50, and so on. For all the rt's and lt's, use multiples of 30—30, 60, 90, 120, and so on.

Using students' suggestions for the fd distance, move the turtle to the first turn in Feed the Turtle. Ask students to estimate the command to turn the turtle the correct amount. Some students may want to use the transparent Turtle Turner. As the turtle executes the turn command, explain what happens on the screen.

Here's a good hint to know how much to turn the turtle: Try to move the turtle into the very middle of each intersection before turning. To help you, the turtle draws a line pointing to where it is heading when it starts each turn. When it turns, it draws a line every time it turns 30°.

Activity

Off-Computer Choices: Estimating Turns

Four Choices   Introduce the Off-Computer Choices. For Sessions 1 and 2, students may choose from three or four choices (Choice 4 is optional). Materials needed for each choice are:

> Choice 1: Turning the Turtle—Student Sheet 7; Turtle Turner
> Choice 2: Turn Commands—Student Sheet 8; Turtle Turner
> Choice 3: As the World Turns—Student Sheet 9; Turtle Turner
> Choice 4: Feed the Turtle Commands—Student Sheet 10 (optional)

Off-Computer Choice 1: Turning the Turtle

Demonstrate how to do Student Sheet 7, Turning the Turtle, by drawing a turn on the board or using a transparency of the student sheet.

For each example, pretend the turtle just drew the solid straight path. The picture shows which way the turtle is facing. I want it to draw this

dotted line next. What should I command it to do next? Which way should it turn? How many degrees?

Ask students first to guess. Encourage students to use their bodies to sweep out the angle through which the turtle turns to help them guess the turn. They then measure each turn using the transparent Turtle Turner. Extending the line in the direction in which the turtle is heading may help them measure the turn accurately when they use the Turtle Turner.

Off-Computer Choice 2: Turn Commands

Explain Student Sheet 8, Turn Commands. Read the directions aloud and use the example on the student sheet, asking a student to walk a path following the commands, then drawing a line on the overhead to illustrate fd 3, demonstrating how to use the Turtle Turner to find rt 60, and then completing the path.

Off-Computer Choice 3: As the World Turns

Explain Student Sheet 9, As the World Turns. Do the first few items together. Discuss how students might use the Turtle Turner on these tasks. Tell students to discuss their ideas with their partners. Some of the items will be done at home as homework.

❖ **Tip for the Linguistically Diverse Classroom** Use pictures and demonstrations to make each direction on this worksheet comprehensible to limited English proficient students. As you do so, have these students draw whatever pictures they find necessary (for example, scissors, toothbrush) to help them recall what they are supposed to do in the direction following each number.

Off-Computer Choice 4: Feed the Turtle Commands (optional)

Some students may want to plan their commands for Feed the Turtle on Student Sheet 10, Feed the Turtle Commands. A small ruler is included on the sheet so students may work on their plans at home if they choose. The ruler is scaled to measure the turtle steps on the computer. To complete the sheet at home, students will need to take home a Turtle Turner.

Classroom Management

After you introduce the On-Computer Activity and Off-Computer Choices, divide the class into two groups and assign one group to work on the computers and the other to do the Off-Computer Choices. Students in each group work in pairs. Switch the groups halfway through the work period. The work period may be quite short in Session 1 since the introductory activities with the whole group take so much time. Therefore, it is probably better to have each group work a short period (for example, ten minutes) on the computer during this session. That way, students in both groups will have experience writing a few commands for Feed the Turtle, rather than students in only one group having the opportunity to work on the computer activity.

Before switching groups, remind students who are using the computers to save their work with a name that includes the students' initials, the date, and the activity (for example, "AR & JA turtle 10/3"). Since students will not have completed Feed the Turtle, tell them they will have more time during the next two days to work on it.

For Session 2, students continue to work in pairs on the On-Computer Activity and the Off-Computer Choices. After about 20 minutes, switch the groups. During the last 20 minutes of Session 2, as a whole-class activity, students share their work by discussing As the World Turns as well as other activities.

Observing the Students As you observe students, ask them to describe their work. For example:

Describe the path you made on the way to the toy.

How many line segments does this path have?

How many corners?

How did you figure out how far to go forward?

How did you figure out what size turns to use?

Show me where the turtle is turning along its path.

See the **Teacher Note**, Turns and Angles (p. 39).

Encourage students to use easily pictured measures, such as multiples of 90, as reference points for estimating amounts of turn. "That looks like 90 and about a third of 90 more. Let's try 120." See also the **Dialogue Box**, Getting a Feel for Degrees and Turns (p. 40).

Sharing Your Work

During the last 20 minutes of Session 2, have students share some of their work. Review, for example, Student Sheet 9, As the World Turns:

How did you measure the turns?

For which turns did you all get about the same answer?

For which turns were there many different answers?

Which turns were particularly hard for you to measure? Why?

Sessions 1 and 2 Follow-Up

🏠 Homework

As the World Turns Students finish Student Sheet 9, As the World Turns, at home. You may also wish to have students do some of their planning for Feed the Turtle at home using Student Sheet 10. They will need to take home a Turtle Turner to do either of these activities.

⬕ Extensions

Turtle Tells Small groups of students play Turtle Tells. One student gives the others commands involving steps and stating turns using degrees; the others follow the commands.

Magazine Turns Have students cut out from magazines and newspapers pictures that contain angles. Have the class group these angles on a large posterboard according to whether the turns required to draw the paths of the angles are 90°, less than 90°, or more than 90°.

Drawing Paths Show students a list of commands such as fd 50, rt 60, fd 50. Students draw the path that would result. Students may want to use the Turtle Turner to help them draw the paths. Then enter the commands on the computer, choosing the Free Explore activity, and compare the drawings to the turtle's path. Repeat the activity keeping the fd 50's the same, but changing the turn command, sometimes putting in a turn command first (for example, rt 90, fd 50, rt 120, fd 50) so students experience different orientations.

Turns and Angles

Understanding angles and angle measures is critical to understanding geometric shapes like triangles and squares. Turtle turning is a powerful and dynamic way to learn about these concepts. In order to take advantage of the potential of turtle turning, students must understand just how turtle turns relate to the shapes the turtle draws.

When students are using Turtle Turners, they need to understand the relationship between the angle that the turtle turns and the angle that is formed when the turtle moves forward in its new direction. For example, the following picture shows the turtle's position after starting on the left, then moving forward 100 toward the right.

fd 100

The next picture shows the position of the turtle and the new direction it is facing after it turns 120°.

lt 120

Below are the results of the turtle moving forward 100 in the new direction.

fd 100

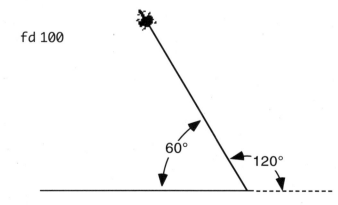

Note that when the turtle moves forward after turning, the angle that it draws is a 60° angle. Even though the turtle has turned through 120°, the lines it draws form a 60° angle.

In grade 3, we are concerned mainly with the measure of the angle the turtle has turned (120°), not with the measure of the angle that is drawn (60°), but in later grades students will work with both. It is important for students to note this difference at the beginning of their work with Turtle Turners.

Notice that with 90° turns, the amount the turtle turns and the measure of the drawn angle are the same:

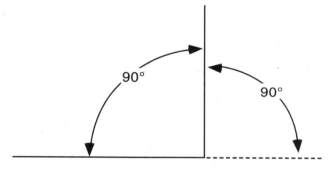

To ensure students are developing sound concepts, you might find it useful to continually check they are visualizing and representing the correct angle through which the turtle turns—120° in the first set of pictures above.

When discussing turns with students, ask them to sweep out the turn with their hands and then turn their bodies. Then ask them to do the same thing in explaining how they are thinking.

Even though we are not concerned with the measure of the angle the turtle draws in turtle turning, students might come to see that the more the turtle turns, the *smaller* the angle produced by drawing the next line will be.

A final historical note: The Egyptians and Babylonians may have divided the circle into 360 equal parts because the year was believed to have 360 days: one degree for one day of the sun's path around the earth.

Getting a Feel for Degrees and Turns

This discussion took place as students did the Off-Computer Choice: Turning the Turtle (p. 35). The teacher noticed that Maya was laying a pencil on her student sheet.

Maya, what strategy are you using there?

Maya: See these stripes on the pencil? I'm using those to find out how much to turn?

How does that help you?

Maya: I just count the stripes. See?—10, 20, 30.

Let me try it. [*She purposely places the pencil farther from the turtle.*] **10, 20, 30, 40, 50 . . .**

Maya: No! That's too much. You did it wrong. Hold it closer.

[*Putting it very close*] **How close? Here?**

Maya: I guess it doesn't work. The stripes don't measure it well. They worked for forwards! Where's the Turtle Turner?

Why won't they work as well for rights?

Dylan: Because those are turns. And turns are in a circle. See those lines on the Turtle Turner, Maya? They go out from the center. They always show the right turn.

The teacher left them talking. She believed Maya was on the way but might need additional experiences getting a sense of turns and their measures.

Earlier, when the pair was doing Feed the Turtle on the computer, they tried typing small increments, such as lt 10 lt 10 lt 10 rt 5 for a 60° turn. The teacher hoped the game would lead them to get a firmer number sense with degrees, because it requires turns of multiples of 30° and does not allow for successive approximation. So she was pleased when she heard the following the next day:

Dylan: Let's turn . . . less than 90, but only a little bit: lt 60!

Maya: We went lt 60. Do a rt 60.

Dylan: Why?

Maya: Because it will turn it back to where it was. Where it was heading before.

Turns, Turtles, and Triangles

What Happens

In Session 3, students define triangles and use their definition to determine whether various figures are triangles. They are introduced to one Off-Computer Choice, Tricky Triangles, which all students complete. Then students may complete the Off-Computer Choices introduced in Session 1. Students working at the computer finish the On-Computer Activity, Feed the Turtle. Their work focuses on:

■ discussing properties of triangles—closed figures with three straight sides and three corners

■ building a definition of triangles from examples and applying the definition to new figures

■ estimating turn measures

The chart below shows how students work during this session.

Materials

■ Choice Time materials from Sessions 1 and 2

■ Student Sheet 11 (1 per student)

■ Student Sheet 12 (1 per student)

■ Student Sheet 13 (1 per student, homework)

■ Computers

Session 3		
Whole Class *20 min*	**Describing Triangles** (Introduction) **Off-Computer Choice: Tricky Triangles** (Introduction)	
Two Groups (working in pairs) *20 min*	Group A **On-Computer Activity** ■ Feed the Turtle	Group B **Off-Computer Choices** ■ Tricky Triangles ■ Turning the Turtle ■ Turn Commands ■ As the World Turns ■ Feed the Turtle Commands
Switch *20 min*	**Off-Computer Choices**	**On-Computer Activity**

Describing Triangles

If we use only the *Geo-Logo* commands fd, bk, lt 90, and rt 90, what shapes could we make? What shapes couldn't we make?

What other commands might we need in order to draw a triangle?

What is a triangle? Can you find any in the room? Can you walk a triangle?

Ask students to describe a triangle in their own words. As they do, draw nontriangles that meet the descriptions. Encourage students to join you in coming up with examples of nontriangles. Take as much time as necessary to have students construct their own definition for the idea of triangle. See the **Dialogue Box**, Putting It in Words (p. 44).

Explain that mathematicians have agreed that triangles are closed paths with three straight line sides joined at three corners. Relate the word *triangle* to other *tri* words, such as *tricycle*, *triple*, and *triceratops* (a dinosaur with three horns).

Off-Computer Choice: Tricky Triangles

Show Student Sheet 11, Tricky Triangles, and Student Sheet 12, Which Are Triangles? (includes directions and space for recording).

I want to see if I can trick you. When you are not working on the computer today, do the Tricky Triangles activity. You can finish the student sheets you worked on yesterday after you finish the Tricky Triangles activity.

Distribute Student Sheets 11 and 12 and explain the directions.

Suggest to students that they ask themselves if the shape has all the properties of triangles. For example:

Is it a closed shape? Does it have three straight sides ? Does it have three corners or angles?

❖ **Tip for the Linguistically Diverse Classroom** Pair limited English proficient students with English proficient students to complete these worksheets.

Classroom Management

Divide the class into two groups and have each group work in pairs. Assign one group of students to finish (or repeat) the On-Computer Activity, Feed the Turtle, and the other group to work on the Off-Computer Choices. During this session, all students need to do the Off-Computer Choice, Tricky Triangles. Those who finish early and have time may do other Off-Computer Choices from Sessions 1 and 2. Switch groups halfway through the work period.

Session 3 Follow-Up

Triangle Cat Give each student a copy of Student Sheet 13, Triangle Cat. Explain that a person drew only triangle paths of different shapes and sizes to make the cat. Read the directions. Discuss strategies for finding triangle paths that are exactly the same shape and same size—that is, congruent.

 **Homework**

Triangle Drawings Students make a drawing entirely of different shape and size triangles.

 **Extension**

DIALOGUE BOX

Putting It in Words

This discussion took place as students did the activity, Describing Triangles (p. 42). The teacher challenged students to describe a triangle in their own words. They had a difficult time verbalizing what they knew and began to draw pictures. When the teacher said not to draw, some formed a triangle with three pencils!

Let's pretend you're talking on the telephone to someone who does not know about triangles. How would you describe a triangle then?

Su-Mei: A triangle has three lines.

So I'll draw such a triangle.

Su-Mei: No! The lines are too far apart! They have to be close together.

Oh, like this?

Su-Mei: No, so close they are touching.

Oh, OK. How about this?

Michael: No. It's not touching right.

Saloni: It's OK. It is a triangle! See, right there.

Several students agreed. But they were finally convinced by the following argument.

Maya: I see the triangle you're talking about. But we're talking about the whole figure! And the whole thing is not a triangle! It's not closed. It has extra lines. If that's a triangle, you could have thousands of lines and call that a triangle, too.

Kate: I know! It's got to be three lines that are connected together and all closed up.

Does everyone agree? If you draw a figure with three lines that make a closed path, does it always make a triangle? Can anyone else help me here?

Ricardo: That's pretty good. But I find something that I don't think is a triangle. Look at this.

Kate: Not that! It has to have three straight lines.

Latisha: I don't think a triangle needs straight lines. It looks like a triangle anyway.

Michael: You can't start curving them. Then you could curve them right in a circle or something.

Kate: I say it has to have three straight lines, all connected.

Can anyone think of a figure that meets that description that is not a triangle?

Samir: Yes. Here's one. He forgot to say a closed path! So I say this as my definition: A closed path with three lines.

Tamara: I have one: A triangle is a closed path with only three lines.

Can anyone think of a figure that meets that description that is not a triangle? Try it.

No one could, and the class agreed that this was a good description. Still, the teacher thought that further discussion was warranted in the future, as several students probably believed that extra segments and curved lines were acceptable.

Equilateral Triangles

What Happens

In Session 4, students discuss Tricky Triangles, applying their definitions of triangles. Students are introduced to an On-Computer Activity, Triangles. They identify equilateral triangles and write *Geo-Logo* procedures for drawing them. As an assessment, they write about triangles. Their work focuses on:

■ discussing their definitions of triangles

■ identifying properties of equilateral triangles—the sides of equilateral triangles are equal in length, and their turns are equal in measure

■ using Logo commands to draw equilateral triangles, estimating turn measures, and using trial-and-error strategies

The chart below shows how students work during this session.

Materials

■ Choice Time materials from Sessions 1–3

■ Transparency of Student Sheet 11

■ Student Sheet 14 (1 per student)

■ Overhead projector

■ Turtle Turner (1 per student)

■ Computers

Session 4		
Whole Class *20 min*	**Discussing Triangles** **Off-Computer Activity: Triangles** (Introduction)	
Two Groups (working in pairs) *20 min*	Group A **On-Computer Activity** ■ Triangles	Group B **Off-Computer Choices** ■ Writing About Triangles ■ Other Off-Computer Choices
Switch *20 min*	**Off-Computer Choices**	**On-Computer Activity**

Activity

Discussing Triangles

Show a transparency of Student Sheet 11, and have students explain their reasoning for determining which Tricky Triangles figures are triangles.

What strategy did you use to be able to know something is a triangle for sure?

If there are disagreements, have students discuss their rationales for identifying triangles. See the **Teacher Note**, Building Strong Concepts (p. 50).

Summarize that to be a triangle, a shape must have all the properties of a triangle: It must be closed, have three straight sides, and have three corners or angles.

If you were explaining to friends exactly how they would decide if a shape were a triangle, what would you tell them to ask themselves about the shape?

Defining Equilateral Triangles   Highlight those triangles on the Tricky Triangles Student Sheet that are equilateral triangles (*C* and *H*) and ask:

What is special about the triangles I marked?

Have students describe the unique properties of the equilateral triangles in their own words. Students may notice that all the sides have the same length and all the turns around the corners are the same size.

Triangles with these properties are called *equilateral triangles*. What do you think *equi-* means in that word?

Lateral refers to side. So *equilateral* means "equal sides."

Draw a shape. Ask students whether the shape is an equilateral triangle or not and have them justify their answers. Suggest to students that they ask themselves if the shape has all the properties of equilateral triangles—for example: Is it a closed shape? Does it have three straight sides of equal length? Does it have three equal corners or angles made by three equal turns?

Activity

On-Computer Activity: Triangles

How might you draw an equilateral triangle in *Geo-Logo*?

Emphasize that knowledge of turns other than 90° turns is necessary to draw many shapes, such as equilateral triangles.

What turn should we use?

In your head, visualize the turtle drawing the triangle.

Would the turn at each corner be 90°, more than 90°, or less than 90°? Why?

Have students turn their bodies to help answer these questions. See the **Teacher Note**, Turns and Angles (p. 39).

Gather students around a computer. Open Triangles. Review *Geo-Logo* commands and the **Erase One tool** .

Ask the class to help you figure out how to write a procedure that draws an equilateral triangle, entering commands as suggested by students. Emphasize the basic decision: Is the turn I used too large or too small? It may help students to discuss that the greater the turn is, the sharper the corner will be. Questions such as the following may be helpful:

How much is the turtle turning? (See the **Teacher Note**, Turns and Angles, p. 39.)

How much do you think the turtle needs to turn?

Did we turn too much or too little?

If students need hints for organizing their thinking, you may wish to introduce a strategy of recording their trial-and-error work. See the **Dialogue Box**, Strategies for Finding the Amount of Turn (p. 52).

When the triangle has been completed, ask questions such as the following:

How many forwards did we use to make the triangle? Why?
How many turns did we use? Why?
How did we figure out the size of each turn?
What part of the triangle does this command draw?

Note: If the students' first command is fd __, the triangle that is drawn will not have a horizontal base. You may need to discuss this if some students think this is not OK.

Tell students that as they work with the computer activity, Triangles, there are some questions they may wish to investigate, such as the following:

What could we change to make a bigger (smaller) triangle? In this activity and other computer activities we will be doing in *Turtle Paths*, you can make paths of any length—they don't have to be multiples of ten.

What is the biggest (smallest) equilateral triangle we could make? (See the **Dialogue Box**, Expanding Your Thinking, p. 53.)

What would happen if we changed all our right turns to left turns?

Can we make a triangle that has a total path length, or perimeter, twice as large as our first one?

Classroom Management

Divide the class into two groups. Assign one group to work on the computer and the other to work at their seats. Switch the groups halfway through the work period.

Students working at the computers make equilateral triangles using the On-Computer Activity, Triangles. Students working at their seats do the assessment activity, Writing About Triangles. All students are to do the assessment. You may wish to have students work individually, rather than in pairs, to do the assessment. If students finish the assessment and have extra time, they can work on the Off-Computer Choices from Sessions 1–3 or explore the questions raised in the extensions.

Activity

Assessment
Off-Computer Choice: Writing About Triangles

Tell students that a new assignment for them to complete is given on Student Sheet 14, Writing About Triangles.

❖ **Tip for the Linguistically Diverse Classroom**   For Question 1, have limited English proficient students circle the part(s) of each shape (for example, where it curves, where it is open) that lead them to conclude it is not a triangle. Offer these students the option of responding to Question 2 in their native languages.

This task provides a window into students' understanding of mathematics and their ability to communicate about mathematics.

Question 1 is similar to Tricky Triangles. Even though the class has discussed triangles during the last two sessions, you may find that some students still do not have a clear understanding of triangles. Some students may be overgeneralizing (believing, for example, that shapes with curved sides, such as #3, can be triangles) or undergeneralizing (believing that #1 is not a triangle "because it's upside-down"). Even those who correctly identify a triangle and a nontriangle may have difficulty writing —or even thinking and talking—clearly about what they know and why they classify triangles the way they do. Both these abilities—to express ideas verbally and to visualize spatially—are important to the development of your students' mathematical thinking.

For Question 2, students may show different ways of thinking about triangles. Some describe parts of a triangle, but not the *process* of constructing one. For example, "It has 3 sides." Others describe the parts of a triangle and a process for putting them together, though their descriptions may be incomplete. "You go up slanty to make one line, then slanty for the next line, then straight for the last line." Others write complete descriptions of parts, properties, and processes—when you follow their directions, you definitely get a triangle!

Students may differ similarly on Question 3. Some give a pattern of *Geo-Logo* commands that make sense for a triangle (for example, fd rt fd rt fd), but without reasonable inputs to those commands. For example, they might write fd 1 rt 2 fd 3 rt 4 fd 5. Others write a basically correct sequence of commands whose inputs are reasonable but not completely correct (for example, with turns that will not work, fd 80 rt 90 fd 80 rt 90 fd 70 rt 60). Some may be able to give a perfect *Geo-Logo* procedure (fd 100 rt 120 fd 100 rt 120 fd 100 rt 120).

Finally, some students perform better on Question 2, with their own natural descriptions. Others express themselves better with *Geo-Logo*, which has given them a language and conceptual framework for thinking and talking about shape and geometry.

Session 4 Follow-Up

 **Extensions**

Closed Paths Try to make a closed path with three, four, five, and six sides (all line segments). Try to make one with one or two sides. Can you make a closed path with three, four, five, and six corners? How about a closed path with one or two corners? (impossible)

Triangles and Almost Triangles Have students draw several figures—some triangles, some closed but not triangles. Have them write a brief explanation for each one.

In this unit, students are asked to invent their own definitions. Wouldn't it be more efficient to just give them the correct definition? Usually not, for several reasons.

First, mathematics is not just "knowing" the definition. Real mathematical activity includes forming and arguing about definitions. Second, students do not think with definitions. They use "concept images"—a combination of all the mental pictures and ideas they have associated with the concept.

For example, students who see only "typical" triangles may say that figures are not triangles

if they do not have a horizontal base ( ) or if they are "long and skinny" ( ).

If students learn only standard verbal descriptions or definitions of a concept, these limited concept images tend to "rule" their thinking. Therefore, students need to build a meaningful synthesis of their own verbal descriptions and a wide variety of visual examples.

For example, consider this class discussion of various Tricky Triangles (Student Sheet 11).

Most students agreed that figure *N* was not a triangle. Ryan, however, thought it was indeed OK because it was two triangles. The class disagreed, convincing Ryan by emphasizing that "it crossed itself" and it "really has four sides."

Opinion was divided on figure *O*. "That's not a triangle! It's too skinny." "Yes, it's just too far up! The lines are going too close in the same direction." "If you stood the two tall sides straight up, then the bottom side wouldn't be straight (horizontal)."

"It looks like the triangle that our large flip-chart makes with the desk, so it must be a triangle."

Finally, one student point-by-point showed how figure *O* fit her definition. "It is too a triangle. It's a closed path, and it's got three straight sides and three corners." Most (but not all) agreed that it was a triangle.

Similarly, figure *P* was rejected on the basis that "it's not closed; no matter how it looks, it's not closed!"

What should you do if students claim that a triangle can have curved sides, or if they start to argue that figure *A* is not a triangle because it doesn't have two "tilted" sides that are equal in length?

Seung: I thought *A* was a triangle, but now I don't think it is anymore!

Why?

Seung: Well, just a minute ago I said that figure *O* wasn't a triangle because it didn't have the two slanted sides that are equal and one flat side. So, *A* couldn't be one either!

One approach is to ask for other opinions, without evaluating any of them. Then ask students to defend their opinions. Usually, the resulting class discussion is more fruitful than students hearing the "correct" definition.

The teacher encouraged Dominic to give an opposing viewpoint when one person claimed that figure *F* was not a triangle because it was upside down.

Dominic: This is a triangle, even though you said it was upside down, because . . . what's that word for same size and same shape.

Congruent?

Dominic: Yes, because, if you turn it over, it would still be congruent.

Continued on next page

To what?

Dominic: To this triangle [*turning the triangle until it is oriented in a typical fashion, with a horizontal base*]—to itself! So, if it's a triangle one time, it must be all the time.

Nevertheless, some students were unconvinced.

The teacher left the issue unresolved for two reasons. First, she believed that respecting and valuing their ideas led to their intense involvement. If she provided a "right" answer, this involvement would be lessened, and students might learn to sit back and wait to be told. Further, some powerful ideas would be sub-

merged. For example, one girl argued strongly that only equilateral triangles were triangles, stating of figure *A*, "It can't be a triangle, because you can't turn it a bit and make it fit on itself. With real triangles, you can." Though she was limiting her definition of triangle more than tradition would have it, her use of *rotational symmetry* to argue her case become a valuable contribution to the discussion.

Second, the teacher believed that students would continue to think about the questions and another discussion the following day would be equally fruitful.

Strategies for Finding the Amount of Turn

The teacher is reading two students' first attempt to define a *Geo-Logo* triangle before going to the computer.

```
to triangle
fd 30
rt 90
fd 30
rt 90
fd 30
rt 90
end
```

Can you draw what the computer will draw?

Students draw what they know. While the first turn is just a bit more than 90°, the second one is much more, so as to make a triangle.

That's not a square corner, is it?

Maya: No, I'll draw it again. [*She draws both as 90° angles, then curves the sides until the figure closes.*]

Maya: [*sounding unsure of herself*] Well, I guess that would work.

Fortunately, *Geo-Logo* provides feedback that is more constrained—the turtle does *exactly* what you tell it to.

At the computer, Maya enters the fd 30 command, then rt 60, rt 10, rt 20. Only then does she figure out that the turtle has turned 90°. She continues to draw a path with three sides, but turns of 90°, 100°, and 110° don't close.

Students debate whether to try 120 or 130. They try both, and they agree that 120 was correct. Other students build images of turns earlier in the process.

Samir: [*putting a hand up, rotating it slowly*] It's 90 . . . [*turning hand a bit more*] 100!

Dominic: It didn't turn enough! See, we need a sharper corner, so we need more of a sharp turn.

Samir: Wait! [*He draws the first side and knows it isn't enough. He tries 110, then 120, and is satisfied.*]

How do you know that the sides are all the same length?

Samir: If you bring this line up, it would be the same.

How else?

Samir: I used forward 40 for all the sides.

A few students seem lost. The teacher decides to get them started by suggesting they use a table to guess and check. That is, students first guess the amount of turn and record their guess on a table. Then they check their guess by trying it in the procedure, and record whether it was too large or too small. Based on this information, they make another guess.

Turn		
Estimate	Too Large	Too Small

Expanding Your Thinking

Help students learn to pursue their own mathematical problems. Often these occur serendipitously. One teacher discussed the procedures they had created together for equilateral triangles and discussed how the numbers changed for making bigger and smaller triangles.

Great. We got the turtle to draw bigger and smaller equilateral triangles. Who can summarize how we did it?

Ryan: We changed all the forward numbers to a different number—but all the same. But the turns had to stay 120, because they're all the same in equilateral triangles.

Elena: We didn't make the biggest triangle.

What do you mean?

Elena: What's the biggest one you could make?

What do you think?

Maya: Let's try 300.

The class tries it.

Rashas: It didn't fit on the screen. All we see is an angle.

Where's the rest?

Dominic: Off here [*gesturing*].

Dominic: Let's try 900!

The student typing makes a mistake and changes the commands to fd 3900.

Whoa! Keep it! Before you try it, tell me what it will look like!

Kate: It'll be bigger. Totally off the screen! You won't see it at all!

Ryan: No, two lines will still be there, but they'll be way far apart.

The students are surprised when it actually turns out the same!

Is that what you predicted?

Ryan: No! We made a mistake.

Rashad: Oh, I get it. It's right. It's just farther off the screen. See, it goes way off, there, like about past the ceiling.

The teacher challenges them to explore this and other problems they can think of.

Samir: I'm going to find the smallest equilateral triangle.

Maya: We're going to try to get all the sizes inside of one another.

Missing Measures

Materials

- Student Sheet 15 (1 per student)
- Student Sheet 16 (1 per student)
- Computers
- Student Sheet 17 (1 per student, homework)

What Happens

In Session 5, students find missing lengths and turns needed to complete partially drawn figures. Students are introduced to the On-Computer Activity, Finding Missing Measures, and the Off-Computer Choice, Missing Lengths and Turns. All students begin work on the Off-Computer Choice, Missing Lengths and Turns. Students work in pairs. As pairs complete their plan for one or two figures, they enter the commands for those figures on the computer.

In Session 6, students continue their work on the Off-Computer Choice, Missing Lengths and Turns, and the On-Computer Activity, Finding Missing Measures. They are introduced to the Off-Computer Choice, Help Make Toys. Session 6 ends with a whole-class discussion about missing measures. Their work focuses on:

- analyzing geometric situations
- applying mathematical processes, such as quantitative reasoning, mental arithmetic, and logic, to find missing measures of figures
- understanding how *Geo-Logo* commands and measures relate to the properties of geometric figures

The following charts show how students work during these sessions.

Session 5	
Whole Class *20 min*	**On-Computer Activity: Finding Missing Measures** (Introduction) **Off-Computer Choice:** ■ Missing Lengths and Turns
Students (working in pairs) *40 min*	All students begin **Off-Computer Choice, Missing Lengths and Turns.** As pairs complete the plans for one or two figures, they go to the computer and enter the commands for only those figures they have completed on paper. They return to their seats to complete other plans, while other pairs use the computers. **Teacher Checkpoint: Missing Measures**

Two Groups (working in pairs) *20 min*	Group A **On-Computer Activity** ■ Finding Missing Measures	Group B **Off-Computer Choices** ■ Missing Lengths and Turns ■ Help Make Toys
Switch *20 min*	**Off-Computer Choices**	**On-Computer Activity**
Whole Class *20 min*	**Whole-Class Discussion: Missing Measures**	

On-Computer Activity: Finding Missing Measures

Today we're going to be turtle detectives. Someone has found some plans for drawing different pictures, but the plans are not finished. We need to use the clues that are there to write procedures so the turtle can draw each picture. Here is the first one. This is supposed to be a rectangle.

Draw the following figure on the board or on a transparency.

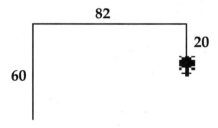

Have students suggest commands to draw the rectangle. Ask them to give reasons for the commands. Reach a consensus. Write the suggested commands beside the drawing.

Gather students around the computer and open Missing Measures. Enter the commands the class wrote. At the end of the procedure, add the command ht to hide the turtle and see that the figure is precisely closed. (The command st shows the turtle.)

Show students how to use the **Label Lengths tool** ![50] to label the length of each line segment on the drawing.

Discuss properties of rectangles that students notice in the drawing and the commands, such as opposite sides are equal in length.

Show students how to click on the **Label Lengths tool**  again to turn it off.

Use the **Teach tool** to teach the procedure to the turtle. Then enter the name of the procedure in the Command Center to see the procedure run.

Show students how they can make changes to the procedure in the Teach window. For example, change each fd 82 to fd 100.

Before clicking out of the Teach window, ask students to predict how the drawing will change. Will the drawing still be a rectangle?

Click in the Command Center and the procedure will be run again automatically, showing the changes.

Activity

Off-Computer Choices: Completing Shapes

Two Choices In Session 5, introduce students to the Off-Computer Choice, Missing Lengths and Turns. All students work on this choice. In Session 6, introduce students to the Off-Computer Choice, Help Make Toys. Student Sheets needed for the choices are:

Choice 1: Missing Lengths and Turns—Student Sheet 15

Choice 2: Help Make Toys—Student Sheet 16

Off-Computer Choice 1: Missing Lengths and Turns

Distribute Student Sheet 15, Missing Lengths and Turns. Tell students they are to work on the problems with their partners. They are to finish each figure so it is a closed path, label the missing lengths, and then write the entire procedure for each closed figure. Remind them that their responsibility is to do a good job thinking—how many of the figures they do is not important.

Off-Computer Choice 2: Help Make Toys

At the beginning of Session 6, explain Student Sheet 16, Help Make Toys. Tell students that they need to label the length of each line on the drawings. There is enough information on the drawings to figure out all the lengths. When they are finished, they draw some toys on the back, labeling the length of each line.

Classroom Management

Students work in pairs. To cycle students effectively, have some pairs go to the computer after planning only one or two figures on Student Sheet 15. Ask students to double-check their work with others before asking to use the computer. Once everyone agrees the figures will work, the students can go to the computer and enter them into the Missing Measures activity. Students should enter only the figures they have planned on the student sheet. They return to their seats to plan the remaining figures.

Students who finish writing the commands on Student Sheet 15 should work on Student Sheet 16, Help Make Toys.

Teacher Checkpoint

Missing Measures

As students are working on paper at their seats or at the computer, observe the strategies they are using by asking them questions, such as:

How are you getting your answers?

What are you trying to do?

How can you get the turtle to make this part?

How can you figure out the number of turtle steps for this side?

Which sides do you know are the same length?

What command will make this corner?

What part of the path is this command making?

See the **Dialogue Box**, Arithmetic in Geometry (p. 60).

Assess how students are using geometry and arithmetic concepts. For example, do they know that sides directly opposite each other in a rectangle will have the same length? Are students mentally computing missing lengths or using paper and pencil to calculate these?

If some students seems to be just guessing, you might ask other students to explain their strategies. Some students might need to use rulers or manipulatives to help them understand missing measures situations.

Analyzing students' work on Student Sheet 15 can provide additional information.

Activity

Whole-Class Discussion: Missing Measures

At the end of Session 6, call students together for a whole-class discussion. Ask several pairs to describe and justify their solutions to the problems. Discuss any differences. For example, some students may have used fd 30 fd 10 fd 20 for the left vertical line on figure 5 on Student Sheet 15, whereas others used fd 60. Are they both valid? Is there a reason to choose one over the other?

How did you find the missing lengths and turns?

How do you know your answers will work?

Describe a strategy you could use to solve any missing lengths problem.

Point out to students that they worked just like detectives, using clues and figuring out information that was missing.

Sessions 5 and 6 Follow-Up

More Missing Measures   Students do the problems on Student Sheet 17, More Missing Measures, for homework. You may want students to exchange the problems they made up for homework when they bring the student sheets back to class.

 **Homework**

Finish the Figure Game   A pair of students who finish early can play this on-computer game. Using the Free Explore activity, the first player enters the commands to draw part of a figure using fd or bk and only 90° turns. The second player enters commands to make a closed path.

Largest Rectangle   Write commands for the turtle to draw the largest rectangle that can be seen in the Drawing window.

 **Extensions**

Arithmetic in Geometry

During the Teacher Checkpoint (p. 57), the teacher observed the following discussion as students worked on Student Sheet 15, Missing Lengths and Turns. On previous activities several students marked off line segments with hatch marks to determine their lengths. So the new activity was a challenge to them since there were no hatch marks.

Maya: But there's no dots! How do we know how far to make the turtle go?

Dylan: How do you count 20 if you don't know where the 20 is at?

What if I walked two steps here and two over here. It would be the same length. The turtle takes the same number of steps every time you give a command.

Rashad: Why is the part on the bottom 20?

Samir: Because it's small.

Su-Mei: Because the space looks about the same as that side that is 20.

Samir: Because it says 20 right there, and it looks like 20.

Maya: This side on the left would be 20, too, because it's the same length as that one on the right.

Dylan: The top would be 70, because it says it right there.

Maya: Seventy for the bottom, too, even though it is unfinished.

Michael: 20 plus 50 is equal to 70.

Maya: You need another 50 on that last side. Because the sides of a rectangle need to be the same length.

Dylan: Just do fd 70. That will do the 20 and the 50.

All the sides?

Maya: No, just the sides here and here. . . .

The sides opposite each other?

Maya: Yes.

The teacher sees that some students are making visual estimates and others are analyzing the figure. Overall, they seem to understand. She is a bit surprised, then, when she observes students working independently on the other figures on the student sheet. For example, students have a variety of ideas about number 4.

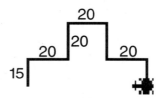

Rashad: [*drawing the bottom line segment*] You can't get this one!

The teacher draws hatch marks to show how the missing part along the bottom could be broken down into three segments.

Rashad: Oh, I see! You just add the top numbers. So, 15 plus 20 plus 20 is equal to 55.

The teacher is taken aback that Rashad just added the first three numbers, from left to right. Samir ignored the existing number labels and other students used different techniques, such as measuring with their fingers. The teacher decides to get students to the computers immediately, so that the feedback will lead them to reflect on their strategies. This helps. For example, Rashad finds his error after entering fd 55.

Continued on next page

After much work and discussion, students begin to be increasingly analytical. Though the teacher knows that many are still working to better understand these tasks, she also knows that all students need to apply arithmetic in problem-solving settings such as these. Later Samir used what he had learned in working on the house figure shown here.

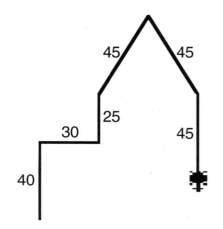

Samir: I thought the right side was 10, because it was such a little space. But now I figured it out: 40, 25, so . . . 65. So, it has to be a 20, not a 10. Because, 20 plus the 45 would be equal to 65.

Paths with the Same Length

What Happens

Sessions 1 and 2: The 200 Steps In Session 1, students work in pairs to form rectangles from straw-string loops. Then they think of procedures to command the *Geo-Logo* turtle to draw rectangles of 200 turtle steps. They are introduced to three Off-Computer Choices and an On-Computer Activity. Then the class is divided into two groups. One group works in pairs on the Off-Computer Choices, and the other works in pairs on the On-Computer Activity. Halfway through the work period the groups switch.

In Session 2, students share their solutions to Student Sheet 18 and the On-Computer Activity. They spend the rest of the work period working on the Off-Computer Choices and the On-Computer Activity.

Sessions 3, 4, and 5: Facing Problems In Session 3, students apply what they have learned in previous sessions to write procedures for shapes of given perimeters that resemble parts of a face. They are introduced to the On-Computer Activity, Geo-Face. Students work in pairs to enter *Geo-Logo* computer commands for a separate procedure for each face part. They are taught how to assemble their face part procedures to form a complete Geo-Face.

In Session 4, pairs work at the computer entering procedures for face parts and assembling the face parts into a Geo-Face. After students complete their Geo-Faces on the computer, they prepare written reports describing and justifying their solutions.

In Session 5, students complete their written reports (if not completed in Session 4) and make presentations of their reports to the whole class. A record of students' work may be created at the end of the session.

Sessions 6 and 7 (Excursion): Designing a *Geo-Logo* Project Students apply what they have learned to design a project of their own creation.

Mathematical Emphasis

- Constructing geometric figures that satisfy given criteria, using analysis of geometric situations, arithmetic, and problem-solving strategies
- Comparing and connecting drawn paths to *Geo-Logo* commands that created them to describe, analyze, and understand geometric figures
- Understanding that shapes can be moved in space without losing their geometric properties
- Estimating and measuring the perimeters of various objects
- Posing and solving original geometric problems

What to Plan Ahead of Time

Materials

- Computers (Sessions 1–7)
- Drinking straws: 3–4 per pair (Sessions 1 and 2)
- String: 30-inch length per pair (Sessions 1 and 2)
- Rulers: 8–10 (Sessions 1 and 2)
- Calculators: 8–10 (Sessions 1 and 2)
- Square tiles: 6 per student (Sessions 3–5, optional)
- Printer (Excursion, Sessions 6 and 7)

Other Preparation

- Duplicate student sheets as follows. If you have Student Activity Booklets, no copying is needed.

  For Sessions 1–2

  Student Sheet 18, 200 Steps and Four 90° Turns (p. 154): 1 per student

  Student Sheet 19, 200-Step Paths That Are *Not* Rectangles (p. 155): 1 per student

  Student Sheet 20, Which Are Rectangles? (p. 156): 1 per student

  Student Sheet 21, Cutting and Combining Rectangles (p. 157): 1 per student (homework)

  For Sessions 3–5

  Student Sheet 22 (pages 1 and 2), Geo-Face Plans (p. 158): 1 per student

  Student Sheet 23, Perimeter Challenges 1 (p. 160): 1 per student (optional)

  Student Sheet 24, Perimeter Challenges 2: (p. 161) 1 per student (optional)

- Work through the following sections of the *Geo-Logo* Teacher Tutorial.

  200 Steps

  How to Make Rectangles in 200 Steps (p. 109)

  More About 200 Steps (p. 110)

  Geo-Face

  How to Make a Geo-Face (p. 111)

  More About Geo-Face (p. 112)

- Plan how to manage the computer activities.

  If you have five to eight computers, have students work in pairs and follow the investigation structure as written.

  If you are using a computer laboratory, it is used in all six sessions of this investigation. You may wish to have students do less extensive planning and checking off the computer, allowing them to refine their plans while working with *Geo-Logo* in the laboratory.

  If you have fewer than five computers, you could cycle students through the computer activities as described in Managing the Computer Activities in This Unit, p. I-21. For Sessions 1 and 2, start students cycling through the computer activity 200 Steps immediately after you introduce it. Other students plan their solutions off the computer as described. Pairs of students continue to cycle through the computer during

Continued on next page

the day, until all students have checked their solutions at the computer. For Sessions 3 and 4, students should plan at least one or two face parts off the computer before entering them on the computer. Other students might plan more. Students cycle through the computers until they have planned and typed each face part. Then they assemble the face during their next turn on the computer.

■ Prepare straw-string loops, one for each pair of students (Session 1). Cut straws into 1-inch lengths. You will need enough straws so each pair has a loop of string with 20 straw pieces. Thread the twenty 1-inch pieces of straw onto a piece of string and tie the ends of the string tightly, like beads, so the straw-string loop will be about 20 inches in total length (perimeter). If the straws are cut ahead of time, your students may be able to prepare the straw-string loops.

■ Decide if you want to change the perimeters used for the Geo-Face activity in Sessions 3 and 4. The first column of numbers we provide below are chosen to present a moderate level of challenge. If you wish to have more of a challenge, use the second column of numbers. (These numbers can be divided by 2, 3, and 4, and so can be used to make rectangles, equilateral triangles, or squares.) If you choose more challenging numbers, the numbers will need to be changed on Student Sheet 22, Geo-Face Plans.

nose	90	72
mouth	200	216
eyes	100 each	96 each
ears	120 each	120 each

The 200 Steps

What Happens

Materials

In Session 1, students work in pairs to form rectangles from straw-string loops. Then they think of procedures to command the *Geo-Logo* turtle to draw rectangles of 200 turtle steps. They are introduced to three Off-Computer Choices and an On-Computer Activity. Then the class is divided into two groups. One group works in pairs on the Off-Computer Choices, and the other works in pairs on the On-Computer Activity. Halfway through the work period the groups switch.

In Session 2, students share their solutions to Student Sheet 18 and the On-Computer Activity. They spend the rest of the work period working on the Off-Computer Choices and On-Computer Activity. Their work focuses on:

- constructing rectangles (and other paths), each having a perimeter of 200 turtle steps, using analysis of a geometric figure, arithmetic, and problem-solving strategies

- finding many different rectangles that have perimeters of 200 turtle steps

- analyzing computer procedures to identify rectangles

The following charts show how students work during these sessions.

- Straw-string loops (1 per pair)
- Rulers (8–10)
- Calculators (8–10)
- Student Sheet 18 (1 per student)
- Student Sheet 19 (1 per student)
- Student Sheet 20 (1 per student)
- Computers
- Student Sheet 21 (1 per student, homework)

Session 1		
Whole Class *20 min*	**Making Paths with 200 Steps** **Off-Computer Choices** (Introduction) ■ 200 Steps and Four 90° Turns ■ 200-Step Paths That Are *Not* Rectangles ■ Which Are Rectangles? **On-Computer Activity: 200 Steps** (Introduction)	
Two Groups (working in pairs) *20 min*	Group A **On-Computer Activity** ■ 200 Steps	Group B **Off-Computer Choices** ■ 200 Steps and Four 90° Turns (done first) ■ 200-Step Paths That Are *Not* Rectangles ■ Which Are Rectangles?
Switch *20 min*	**Off-Computer Choices**	**On-Computer Activity**

Session 2		
Whole Class *20 min*	**Discussing 200 Steps and Four 90° Turns**	
Two Groups (working in pairs) *20 min*	**Group A** **On-Computer Activity** ■ 200 Steps	**Group B** **Off-Computer Choices** ■ 200 Steps and Four 90° Turns ■ 200-Step Paths That Are *Not* Rectangles ■ Which Are Rectangles?
Switch *20 min*	**Off-Computer Choices**	**On-Computer Activity**

 Ten-Minute Math: Lengths and Perimeters Two or three times during the next few sessions (outside of math time), continue the activity Lengths and Perimeters. Try the variation that involves the perimeters of regular polygons. You may need to introduce your students to the idea of perimeter by having students draw shapes on paper and "walk" around them with their fingers or by using tape to make shapes on the floor and having students walk around them. Provide a perimeter problem such as the following:

The turtle was given this command: repeat 4 [fd __ rt 90]. **When it was finished, the turtle had drawn a closed shape with a perimeter of 40 turtle steps. What shape did it make? What was the number for the** fd **command?**

Students work in pairs for two or three minutes to sketch what they think the turtle drew and to mark the lengths of the sides and the perimeter on their sketches. Ask students to contribute their ideas. Students can demonstrate by acting out the turtle commands themselves. For complete directions and variations, see p. 82.

Activity

Making Paths with 200 Steps

Give each pair of students a straw-string loop. Have the pairs work together to make the loop form as many different sizes and shapes of rectangles as possible. Have each group describe one of their rectangles.

What are the lengths of each side? How many steps would it take to walk all the way around your rectangle? The distance all the way around a rectangle (or any other closed shape) is called the *perimeter*.

What is the same about all the rectangles? (Make sure the 90° turns making right angles is discussed.)

What are some differences between the rectangles? Are there any other rectangles that we can make with the straw-string loops? Can other rectangle shapes be formed? How?

Pose the following investigation to the class:

Today we're going to investigate procedures for describing paths of exactly 200 turtle steps. Some paths will be rectangles; others may not be rectangles. Some paths will be closed; others may be open.

Think of a procedure that would make the turtle draw a closed path that has a total length of 200 turtle steps and four 90° turns. The fourth turn takes the turtle back to the heading at which it started. Draw a quick sketch of the turtle's path, labeling the lengths of the lines. Who has some ideas as to how we would write this procedure?

Take suggestions from different students to find a solution. Ask students to describe their procedures and justify why they fit the rule of drawing a closed path of total length—or *perimeter*—of 200, with four 90° turns. You may wish to demonstrate a few procedures on the computer, opening and using the 200 Steps activity. Ask students to describe the shapes that were formed.

Are they all rectangles? Is it possible to make a shape of 200 turtle steps with four 90° turns that is not a rectangle? (See the **Dialogue Box**, Geometric Number Sense, p. 70.)

Off-Computer Choices: 200 Steps and Rectangles

Three Choices During Sessions 1 and 2, students work on three Off-Computer Choices. Students should do Choice 1 before doing the other two choices. They may do some of the tasks on the student sheets, then check their solutions on the computer. Or they may do the task initially on the computer, then enter their solutions on their student sheets and look for other solutions. Students will need the following materials for the choices:

> Choice 1: 200 Steps and Four 90° Turns—Student Sheet 18, string-and-straw loops, rulers, calculators (optional)
>
> Choice 2: 200-Step Paths That Are *Not* Rectangles—Student Sheet 19
>
> Choice 3: Which Are Rectangles?—Student Sheet 20

❖ **Tip for the Linguistically Diverse Classroom** Make sure you read aloud the directions on Student Sheets 18 and 20, modeling actions and drawing pictures whenever necessary to ensure comprehension.

Off-Computer Choice 1: 200 Steps and Four 90° Turns

Explain the instructions for Student Sheet 18, 200 Steps and Four 90° Turns. Tell students they are to do this choice before doing the other two choices. They will be able to check some of their work by testing it on the computer. Students may use the string-and-straw loops, rulers, calculators, and other materials to find possible rectangles that can be made with 200 steps.

Off-Computer Choice 2: 200–Step Paths That Are *Not* Rectangles

Explain Student Sheet 19. Tell students they may check some of their solutions on the computer using the activity, 200 Steps.

Off Computer Choice 3: Which Are Rectangles?

Explain Student Sheet 20. Most students will not need to check their solutions on the computer, although some may wish to do so.

Activity

On-Computer Activity: 200 Steps

Tell students to open the activity, 200 Steps, on the computer and try to make closed paths that have a length of 200 steps and four 90° turns. The fourth turn takes the turtle back to its starting position.

Tell students to use the **Label Lengths tool** to label the length of each line segment on their drawing. Remind students of the ht (hide turtle) command they can use at the end of the procedure so they can see if their path is closed. Have students teach the computer (using the **Teach tool**) any solutions and to save their work when they are done.

Students may also use the 200 Steps activity on the computer to make paths that are not rectangles (from Student Sheet 19).

Classroom Management

Divide the class into two groups. Assign one group to work in pairs on the On-Computer Activity and the other to work in pairs on the Off-Computer Choices. Switch groups halfway through the work period.

Students who use the computers first can try procedures before writing them on Student Sheet 18, 200 Steps and Four 90° Turns. If they are successful, they should copy their procedures onto the student sheet. Students who use the computers after their desk work should test some of their procedures using the 200 Steps computer activity.

If some students need help with computation, show them how to have *Geo-Logo* act as a calculator. For example, if they type pr 85 + 15, *Geo-Logo* will print 100.)

Discussing 200 Steps and Four 90° Turns

At the beginning of Session 2, ask students to share some of their solutions to 200 Steps and Four 90° Turns (Student Sheet 18 and the On-Computer Activity). Help students see how the change in measures affects the change in shape. Pose problems and ask questions.

Were all the shapes you made rectangles? Why or why not? The lengths of the sides of a rectangle are 80, 20, 80, 20. Draw a picture of it on your own paper. First picture in your mind and then draw another rectangle with side lengths of 20, 80, 20, 80. How did your second drawing change? Are the two drawings the same shape or not? In other words, are these two shapes congruent—the same shape and the same size? What is different about them?

Ask students to draw a rectangle with side lengths of: 70, 30, 70, 30.

Is this the same shape as the other two—is it congruent to the other two? How is it different? How is the shape changing? How can you predict that by knowing the measure of the sides?

Discuss the different side lengths students found for rectangles on Student Sheet 18 and how they could find different solutions.

What strategy could you use to change one solution into a different solution that would also work? What could you do to find every solution? Could we invent a pattern or strategy that would let us find every possible rectangle with a perimeter of 200 steps? If two rectangles have exactly the same total path length, or perimeter, must they have the same size and the same shape? Can you convince us you are right?

After the discussion, students return to the On-Computer Activity and Off-Computer Choices.

Sessions 1 and 2 Follow-Up

Cutting and Combining Rectangles   Send home Student Sheet 21, Cutting and Combining Rectangles. Students figure out the perimeter of part of a rectangle or of several rectangles combined into one. Some of these problems may be challenging, because students have to visualize how rectangles might be cut apart or joined together.

 **Homework**

Geometric Number Sense

After introducing the 200 Steps problem, the teacher asked his students to sketch some initial solutions. He noticed several students drew rectangles but labeled the sides with numbers that could not be the measures of a rectangle's sides, such as 70, 30, 60, 40.

He believed that having students see the results of their thinking would be best. So he asked them to try to draw such a figure with *Geo-Logo*.

Maya: It's not a rectangle!

Dominic: Because it's like this [*drawing in the air*]. Because one side was not as long as the other one. So, we have to do that again.

The students change their procedure to a rectangle of sides of 60 and 30 and believe at first that this is 200 steps in total length. With discussion prompted by the teacher, they change it to a rectangle with sides of 70 and 30.

The teacher observes other behaviors that help him understand his students. Most students are not facile with mental computation and do not use computation shortcuts. For example, they check the total length of a rectangle with sides of 70 and 30 by using a written algorithm:

$$
\begin{array}{r}
70 \\
70 \\
30 \\
+30 \\
\hline
\end{array}
$$

They do not "see" or use the 70 + 30 = 100 shortcut.

Several students attempt to make a square with a perimeter of 200. After trying 25 for each side then 75 for each side, they are ready to conclude it is not possible! The teacher plans to work more on number sense in this and other activities.

Toward the end of the activity, the teacher notices that many students are developing more flexibility in thinking.

Samir: First I wanted 10 and 10, then I couldn't get it, so I made it 20, 80, 20, 80.

What if you wanted it 10 and 10?

Samir: Well, then, that would have to be 90 and 90.

How did you do that?

Samir: I just added the 10 on to both sides because you can't leave it out!

Facing Problems

What Happens

In Session 3, students apply what they have learned in previous sessions to write procedures for shapes of given perimeters that resemble parts of a face. They are introduced to the On-Computer Activity, Geo-Face. Students work in pairs to enter *Geo-Logo* computer commands for a separate procedure for each face part. They learn how to assemble their face part procedures to form a complete Geo-Face.

In Session 4, pairs work at the computer, entering procedures for face parts and assembling the face parts into a Geo-Face. After students complete their Geo-Faces at the computer, they prepare written reports describing and justifying their solutions.

In Session 5, students complete their written reports (if not completed in Session 4) and make presentations of their reports to the whole class. A record of students' work may be created at the end of this session. Their work focuses on:

- constructing rectangles, squares, and triangles of a given perimeter
- understanding that shapes can be moved in space without losing their properties; shapes in different locations and orientations may look different but still be congruent
- estimating and measuring the perimeter of various objects

The following charts show how students work during these sessions.

Materials

- Student Sheet 22, pages 1 and 2 (1 per student)
- Student Sheet 23 (1 per student, optional)
- Student Sheet 24 (1 per student, optional)
- Square tiles (6 per student, optional)
- Computers

Session 3	
Whole Class *20 min*	**Facing the Challenge** **On-Computer Activity: Geo-Face** (Introduction)
Students (working in pairs) *40 min*	As pairs complete the plans for one or two face parts, they go to the computer and enter the commands for only those parts they have completed on paper. They return to their seats to complete other plans, while other pairs use the computers.

Session 4		
Whole Class *20 min*	On-Computer Activity: **Geo-Face** (Introduction, continued)	
Two Groups (working in pairs) *20 min*	Group A **On-Computer Activity: Geo-Face** ■ Finish entering face parts and assemble Geo-Face	Group B **Off-Computer Choices** ■ Finish Geo-Face Plans ■ Prepare written reports ■ If reports are finished, do Perimeter Challenges (Extension)
Switch *20 min*	**Off-Computer Choices**	**On-Computer Activity**

Session 5	
Whole Class *60 min*	If necessary, students finish their reports. Presentation by students of their written reports.

Activity

Facing the Challenge

For the next three sessions, we're going to use several things that we've learned in *Geo-Logo* to do something creative. We're going to design a face—a Geo-Face.

Distribute copies of Student Sheet 22, pages 1 and 2, Geo-Face Plans. Read the directions aloud. Work through one example of a face part, such as an ear, with the class.

❖ **Tip for the Linguistically Diverse Classroom** Before limited English proficient students begin Student Sheet 22, make sure they understand the meaning of the word *face* by drawing one on the board. As you point out the nose, mouth, teeth, eyes, eyeballs, and ears on your drawing, have them draw these parts over the corresponding words on the student sheet.

Let's think about how we could make an ear. The directions say that each ear needs to have a perimeter of 120. Could you make a rectangle, or a square, or a triangle with that perimeter? How?

Take suggestions from several students. Discuss how ears can be different shapes or have different dimensions and still have a perimeter of 120.

On the student sheets, students draw their face in the rectangle on page 2 and label the lengths on each face part. Then they are to write commands for each face part as a separate procedure. Tell them that at the end of each procedure it's important *to turn the turtle so it is facing up, in the direction it started*—this will make it easier for them to put the face parts together.

Activity

On-Computer Activity: Geo-Face

Tell students that after they have drawn and written commands for one or two face parts, they enter their face parts on the computer in the activity called Geo-Face. Open Geo-Face.

Students enter each face part as a separate procedure. Demonstrate by entering a face part planned in the previous activity, such as an ear. The face part will be drawn in the middle of the screen. If asked, tell students that they need not make two ears at this time or be concerned that the part is not in the correct place. Demonstrate how to use the **Teach tool** to teach the computer the procedure before entering the next face part. (During Session 3, students can ignore the head procedure they see in the Teach window.)

Tell students that in Session 4 you will show them how to assemble their face procedures to make a whole Geo-Face.

At the beginning of Session 4 or when students have entered most of their face parts onto the computer, gather them around a computer to show them how to assemble their faces. You may want to use the face part procedures written by one pair of students. (If you do, don't save your demonstration work when done so the pair can reassemble their face.)

Open Geo-Face on the computer. Enter head to draw the outline of the face. Drag the turtle to a location where a face part will go. Enter the name for that face part procedure. Drag the turtle to the location for the next face part and so on. (See p. 111 in the *Geo-Logo* Teacher Tutorial.)

Also demonstrate how to use the mouse motions, sliding and turning to move face parts if necessary. (See p. 119 in the *Geo-Logo* Teacher Tutorial.)

Note: Each face part must have been entered and taught as a procedure before it can be slid or turned. For an example of how a face was assembled, see the **Teacher Note**, Assessment: Planning, Assembling, and Presenting Geo-Faces (p. 77).

Adding Color to the Faces If you have color or gray-scale monitors, show students how to add color to the faces with the setc command. Remind them that this should be the last step in building their faces. (See p. 116 in the *Geo-Logo* Teacher Tutorial.)

Classroom Management

It probably will take your students two or three days to plan and assemble their Geo-Faces. During Session 3, begin by having all students, working in pairs, plan only one or two face parts on Student Sheet 22. As each pair of students completes a face part, have them teach the procedures to the turtle on the computer. When they are finished, have them return to their seats to plan additional face parts. Continue cycling through the pairs until all students have had a chance to work at the computer. Remind students to save their work before leaving the computer. During Session 4, you might return to the typical way of working in this unit with half the class using the computers for half the work period. During Session 5, have students finish their written reports (if they didn't finish them in Session 4) and present them to the class.

Activity

Assessment

Planning, Assembling, and Presenting Geo-Faces

Observe students as they plan and assemble their Geo-Faces. You may wish to look for the following:

- Do students approach the problem in a systematic way, planning and completing one face part before doing another?
- Are students able to work within the constraints of the problem, such as making the perimeters match the requirements and including at least one rectangle, square, and triangle?
- Do students use their previous experiences in this unit to help them write commands for different procedures?
- Do students have a systematic way of organizing and recording their procedures so they are clear and easy to enter into the computer?
- Are students able to follow their plans and assemble their faces on the computer?

If you have a printer available, students can print out their face procedures and computer drawings.

When their faces are complete, students describe and justify their solutions in written reports. This should include the procedures they wrote (if they are not on a computer printout) and an explanation of how each one of their face parts "fits the rules." Ask students to expand and clarify their explanations as necessary.

❖ **Tip for the Linguistically Diverse Classroom** Have limited English proficient students describe and justify their solutions by using pictures and numbers.

Students could present their reports to the class. For example, they might run their procedure on the computer and present the written report orally. If available, they might show a transparency of the printout of their *Geo-Logo* procedures (or use a blank transparency with the procedures written on it).

Refer to the **Teacher Note**, Assessment: Planning, Assembling, and Presenting Geo-Faces (p. 77), for commentary on some sample student responses.

Choosing Student Work to Save

As the unit ends, you may want to use one of the following options for creating a record of students' work:

- Students look back through their folders or notebooks and write about what they learned in this unit, what they remember most, and what was hard or easy for them. You might have students do this work during their writing time.

- Students select one or two pieces of their work as their best work, and you choose one or two pieces of their work to be saved. This work is saved in a portfolio for the year. You might include students' written responses to the assessment, Off-Computer Choice: Writing About Triangles (Investigation 2, Session 4), and any other assessment tasks from this unit. Students can create a separate page with brief comments describing each piece of work.

- You may want to send a selection of work home for parents to see. Students write a cover letter, describing their work in this unit. This work should be returned if you are keeping a year-long portfolio of mathematics work for each student.

Homework

Geo-Face Plans   After Session 3 and/or 4, students take home a copy of Student Sheet 22, Geo-Face Plans, so that they can continue to plan parts of their face drawings. Each of the two students in a pair should work on a separate copy of the student sheet (you will need fresh copies for some students); they can combine their work when they return to class the next day. If they have already planned the eyes, ears, nose, and mouth, they can try the challenges: teeth, a hat, earrings, eyeglasses, or other features. When they return to class, they will be able to use the computer to test their procedures and assemble their face.

Extensions

Switching Geo-Face Plans   Have each pair of students plan another Geo-Face and draw it on paper, labeling all the sides and turns. Then have each pair switch their plan with another pair of students and attempt to draw each other's face on the computer with the Geo-Face activity. Have them discuss any problems or discrepancies with the other pair.

Perimeter Challenges   If students have completed their written reports before the end of Session 4, give them Student Sheets 23 and 24, Perimeter Challenges 1 and 2. Students measure the objects in centimeters or inches.

❖ **Tip for the Linguistically Diverse Classroom**   As you point to and identify each object named in the left column of Student Sheet 23, have limited English proficient students draw pictures over the corresponding words. Offer these students the option of responding to numbers 2 and 3 on Student Sheet 23 and number 3 on Student Sheet 24 in their native languages.

Assessment: Planning, Assembling, and Presenting Geo-Faces

Students' planning of their face parts is one of the most substantial mathematical aspects of the assessment activity. Close observation and work with students will provide you with information about what students understand.

Chantelle The teacher observed Chantelle as she worked. Chantelle had written the following procedures.

to nose	to eye	to mouth	to ear
fd 30	fd 25	fd 20	fd 80
rt 120	rt 90	rt 90	rt 90
fd 30	fd 25	fd 80	fd 20
rt 120	rt 90	rt 90	rt 90
fd 30	fd 25	fd 20	fd 80
rt 120	rt 90	rt 90	rt 90
end	fd 25	fd 80	fd 20
	rt 90	rt 90	rt 90
	end	end	end

To assemble her face, she first entered head. She then clicked and dragged the turtle to the location she wanted for the eyes and ears, entering the command for each procedure. But she didn't like the way the second ear (E) looked.

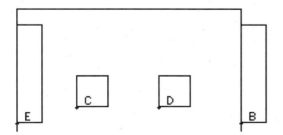

She clicked on the side of that ear and dragged it into position. The command slide –20 0 was automatically added to the Command Center.

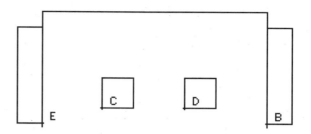

She then dragged the turtle to a position for the nose procedure and entered the nose command, then dragged the turtle to a position for the mouth procedure and entered the mouth command. She dragged the point G down a bit and the mouth was redrawn there.

She still, however, didn't like the nose's heading, or orientation. So she clicked on a corner and dragged it to turn it.

The procedure max was the entire face shown below.

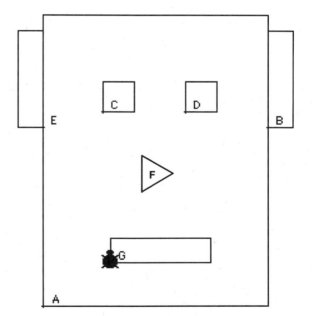

Continued on next page

```
to max
head
make-points [B]
jumpto B
ear
make-points [C]
jumpto C
eye
make-points [D]
jumpto D
eye
make-points [E]
jumpto E
ear
slide −20 0
make-points [F]
jumpto F
nose
make-points [G]
jumpto G
mouth
nose
turnit [rt 30]
ht
hp
end
```

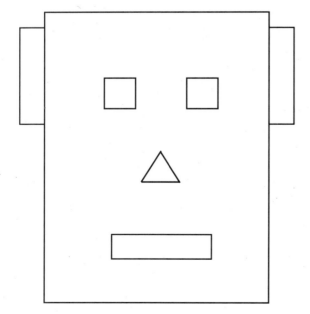

Chantelle's work was complete, organized, and accurate. She could explain to the class why each of her procedures followed the rules.

"My nose has a perimeter of 90 steps because it's an equilateral triangle with 30 for each side. My eye's a square, and 4 times 25 is 100. My mouth is just like the 200 steps rectangles. I cheated on the ear. I figured it was OK to just count *one* long side of 80 and two short sides of 20 to get a total of 120, because the head already drew the last side."

The teacher knew that Chantelle understood perimeter; was facile at the mental computations involved; and understood certain properties of equilateral triangles, squares, and rectangles. Just as important, Chantelle connected her knowledge of number to her knowledge of geometry.

"I made the mouth 20, 80, 20, 80 because that was short and wide; the ears had to be 80, 20, 80, 20 to be tall and thin. I could have picked 60, 40, 60, 40, but I didn't want them to look like squares."

Jeremy Jeremy presented a slightly different picture to his teacher. He tried to use squares for all his face parts. Each ear, for example, was 15 steps on a side. The teacher asked him about that, for the specified perimeter for the ears was 120. "I was going to do 120 for each ear, but I couldn't figure out 120 divided by 4. So I did 120 for both ears . . . 60 for each." Jeremy explained that he eventually used a triangle for the nose because "I could divide 90 by 3 easier."

The teacher asked him how he knew the number for the turns in his nose procedure. "You have to use 120, because you go up, and that much [*gesturing perpendicularly*] would be 90, and you need 120 to go down at this heading."

Continued on next page

Jeremy's teacher understood that he could perform division in his head easily; however, he avoided doing slightly more difficult computations, possibly because that would require paper-and-pencil computation.

He was competent at the task but perhaps needed encouragement to extend his abilities.

Maya Maya drew the shapes she wanted and only later tried to make the numbers fit. Then she didn't ignore scale completely but didn't really adjust her drawings, either. Even though she didn't draw to scale, she was conscious of things fitting inside. She realized the numbers had to be less for the "inside shapes."

Through observations during the planning and programming of her project, the teacher learned that Maya had only a limited number sense before the project, but learned a lot as she planned and developed her face project. For example, Maya found that "71, 30, 71, 30" would not yield a perimeter of 200 for the mouth. She noticed quickly that this would not work, but it would if she made both the 1's into 0's.

She was also a lot more willing to try numbers that were not multiples of 10 than she had been previously. She used a successive approximation strategy for this. For example, if 10 was too small and 20 too big, she would try 15, then 14, 13, 12, and so on. The teacher was pleased to see her simultaneously develop number sense, systematic problem-solving strategies, and knowledge of the properties of geometric figures.

Designing a *Geo-Logo* Project

What Happens

Students apply what they have learned to design a project of their own creation. Their work focuses on:

- analyzing geometric figures in a drawing to create *Geo-Logo* commands that will make the turtle re-create that drawing
- understanding that shapes can be moved in space without losing their properties
- posing and solving one's own mathematical problems

Materials

- Computers
- Printer

Activity

A *Geo-Logo* Project

Have students use their face parts from Sessions 3–5, along with new procedures they plan and write, to draw a design or picture. You may wish to work with students who begin overly complex pictures to ensure they start with something manageable. See the **Teacher Note**, Problem Posing in *Geo-Logo* (p. 81). Plan to devote at least two sessions to this activity.

If time permits, display students' work. Have students help you create the display, including printouts of each of the following for each student or pair of students:

- the picture
- the written procedures that drew the picture
- a written description of the picture and how the problem of drawing the picture was solved

Students can use these printouts in describing their project to the class.

Continuing such projects throughout the year is extremely valuable. Students grow in their ability to analyze and solve problems. See the **Teacher Notes**, Problem Posing in *Geo-Logo* (p. 81), and Styles of Programming (p. 81).

Problem Posing in Geo-Logo

Mathematics is not just getting right answers. It's not even just solving problems. A critical mathematical activity is *posing problems*. Students might do this in a variety of ways.

One valuable way for students to learn to pose problems is to have them create their own *Geo-Logo* pictures or projects.

For example, students have enjoyed rotating their equilateral triangle around the "home" position of the turtle, using the repeat command (or just typing the commands in again and again).

```
to tri
repeat 3 [fd 50 rt 120]
end
repeat 12 [tri rt 30]
```

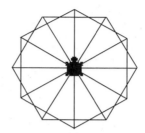

Others enjoy making pictures of buildings or flowers. They might work on such projects during spare moments for a long period of time, even months. This would be a good time to begin such projects. These projects motivate students to apply mathematics. Students develop persistence and pride of creation, as well as learn how to set goals and problems for themselves.

Styles of Programming

Think of Chantelle's work, described in the **Teacher Note**, Assessment: Planning, Assembling, and Presenting Geo-Faces (p. 77). She taught each face part to the computer separately, then eventually combined them. Chantelle used top-down programming. Top-down programmers often prefer to plan in advance, have a clear idea of the end result, and look at the big picture. They may even break down each of the smaller parts, until they have specified every step. One student described these parts as "mind-sized bites." These students often use mathematical analyses.

Bottom-up programmers plan as they go, using what they see happening to make decisions about what to do next. They build up a program piece by piece, discovering what works as they proceed. Often, especially in their early stages of learning, they do not use subprocedures at all. Differences in programming style are often reflected in different strategies for geometric thinking.

You cannot always determine what strategies students have used by looking at their *products* alone. You must understand their *processes*; that is the important aspect. Also, many students use a combination of the two approaches.

Often, the project students choose determines the style of programming they use. Some like to begin with no firm idea of what they wish to draw. These *loosely defined goals* usually lead to bottom-up programming. Others prefer to have a *well-defined goal*, such as an abstract design.

Both types of goals, and both styles of programming, are valuable. Loosely structured goals can lead to beneficial explorations and a "feel" for the geometry of the situation. Well-defined goals can help students analyze situations and grasp the structure of the problem and the geometry. If some students always approach projects in one way, encourage them to work occasionally on types of projects (either more or less defined) they tend not to choose.

Lengths and Perimeters

Lengths and Perimeters is an off-computer problem using the *Geo-Logo* repeat command.

Basic Activity

Students try to visualize the results of a repeat command given to the *Geo-Logo* turtle. First they work with length. Given a total distance (for example, 72), they think of repeat commands that will send the turtle that distance (for example, repeat 6 [fd 12]). By working with these commands, they are working with factors and multiples.

Students also do this activity with shapes. Given a perimeter of a regular polygon, they figure out the repeat command to make that shape. Or given the repeat command, they figure out the perimeter. Other variations involve rectangles.

Students focus on:

- relating factors to their multiples
- recognizing, visualizing, and drawing polygons
- relating the perimeter of a polygon to the lengths of its sides
- using multiplication, division, addition, and subtraction in the context of the perimeters of polygons

Materials

- Calculators (optional)
- Computer with large demonstration screen and the *Geo-Logo* disk for *Turtle Paths* (optional)

Procedure

Note: The repeat command is not formally presented in this unit, although it is available for students to use on the computer. If your students have not used the repeat command, you will need to introduce it the first time you do this activity. For hints, see the Special Notes on the next page.

Step 1. Choose a distance that you want the turtle to go. For example, What could we do so the turtle would go 35 turtle steps?

Step 2. Students write down commands to move the turtle that distance. Working in pairs, students spend two or three minutes writing down all the *Geo-Logo* commands using repeat that they can think of that would send the turtle that distance. For example, for 35 turtle steps, these commands would work:

repeat 5 [fd 7]

repeat 7 [fd 5]

repeat 35 [fd 1]

repeat 1 [fd 35]

Students can use calculators to test their ideas.

Step 3. Make a list on the board or overhead of all the different responses. Ask students how they know that the commands work. Students can use skip counting, demonstrate with concrete materials, or explain their mental strategies to prove their answers. Ask if they have all the possibilities. How do they know? Could 3 work? What about 9? What about 14? How do they know?

Variations

Perimeters of Regular Polygons These problems either provide the perimeter of a shape and students have to find out the length of a side, or they give the length of a side and students have to determine the perimeter. Here are some examples:

- The turtle was given this command: repeat 4 [fd __ rt 90]. When it was finished, the turtle had drawn a closed shape with a perimeter of 40 turtle steps. What shape did it make? What was the number for the fd command?
- The turtle made a regular hexagon—a six-sided shape just like the yellow pattern block. The perimeter of the hexagon was 72 turtle steps. Find the number for the blank in the command it used: repeat 6 [fd __ rt 60].

Continued on next page

- The turtle made a triangle with this command: repeat 3 [fd 35 rt 120]. What is the perimeter of the triangle?

In these problems, encourage students to sketch what they think the turtle drew and to mark on their sketch the lengths of the sides and the perimeter. Ask students to act out what the turtle did by walking its path on the floor.

Perimeters of Other Shapes These problems involve shapes in which all the sides are not equal. Here are some examples:

- The turtle made a rectangle using the following command: repeat 2 [fd 20 rt 90 fd 10 rt 90]. What did the rectangle look like? What is its perimeter?

- The turtle made a rectangle with a perimeter of 50. It made the shape with this command: repeat 2 [fd 12 rt 90 fd __ rt 90]. What is the missing number in the fd command?

- The turtle made a rectangle with a perimeter of 50. It made the shape with this command: repeat 2 [fd __ rt 90 fd __ rt 90]. What are the missing numbers in the two fd commands? Is there more than one set of numbers that will work?

Make sure students sketch what they think the turtle drew and mark on their sketch the lengths of the sides and the perimeter. As you walk around the room, you can easily see from their sketches what students understand and what is confusing for them as they visualize the turtle's movements. You may also want to have students act out the turtle's movements.

Using Decimals Introduce the use of .5 into the problems. For example:

- The turtle was given the following command: repeat 4 [fd 6.5]. How many turtle steps did it take?

Discuss the meaning of .5 (for ways to develop this idea, see the grade 3 unit, *Mathematical Thinking at Grade 3).* Students can try the problem mentally, then test their ideas on the calculator.

Special Notes

The repeat Command in *Geo-Logo* If your students have not yet used the repeat command on the computer, you will have to spend the first ten-minute math session introducing this command.

Write a repeat command on the board—for example, repeat 20 [fd 2]. Explain that the directions inside the bracket are done over and over again, as many times as the number following the word repeat. So repeat 20 [fd 2] means: take two turtle steps (that's the first time), take two turtle steps (that's the second time), take two turtle steps (that's the third time), continue taking two turtle steps until you've done that 20 times.

Have students act out a few different repeat commands. Be sure they try some examples that have more than one command in the brackets. For instance, repeat 4 [fd 40 rt 90] means that the turtle will do the two commands fd 40 rt 90 four times:

fd 40 rt 90 fd 40 rt 90 fd 40 rt 90 fd 40 rt 90

See the *Geo-Logo* Teacher Tutorial for more information on using the repeat command (p. 116).

Perimeter If students are not familiar with the word *perimeter*, explain that it is the distance around the outside of a closed shape. Draw a closed shape on the board and say:

If an ant started here and walked all the way around the shape until it came back to where it started, it would have walked around the shape's perimeter.

Put some closed shapes on the floor with tape and have students walk around their perimeters. Ask them how far they went. They can find solutions in paces or in standard measures (for related activities on linear measure, using nonstandard and standard units, see the grade 3 unit, *From Paces to Feet).*

Continued on next page

Testing Students' Commands on the Computer If a computer with a large demonstration screen is available, you can test students' commands so that all students can see the results. Otherwise, students can test the commands on their own later at the computer. Use the Free Explore activity in *Geo-Logo*.

There are several ways to test students' repeat commands to see if the turtle travels the correct distance.

- First, have the turtle make the line (for example, fd 35). Then bring the turtle back to the beginning of the line and test each of the student's repeat commands to see if they bring the turtle the same distance.

- Draw a line of the correct length (for example, 35 turtle steps). Position the turtle so that it will draw a line parallel to the first line. This lets students more easily compare their lengths.

- Use the **Ruler tool** in *Geo-Logo* to measure the lines made by students' repeat commands. However, keep in mind that it is easy to be off by one or two turtle steps when measuring with the Ruler. When using the Ruler with decimals, you will want to make sure that numbers are displayed using one decimal point (change the number of decimal places by selecting **Decimal Places** from the **Options** menu).

The following activities will help ensure that this unit is comprehensible to students who are acquiring English as a second language. The suggested approach is based on *The Natural Approach: Language Acquisition in the Classroom* by Stephen D. Krashen and Tracy D. Terrell (Alemany Press, 1983). The intent is for second-language learners to acquire new vocabulary in an active, meaningful context.

Note that *acquiring* a word is different from *learning* a word. Depending on their level of proficiency, students may be able to comprehend a word upon hearing it during an investigation, without being able to say it. Other students may be able to use the word orally but not read or write it. The goal is to help students naturally acquire targeted vocabulary at their present level of proficiency.

We suggest using these activities just before the related investigations. The activities can also be led by English-proficient students.

Investigations 1–3

closed, open

1. Draw these two shapes on the board.

2. Show how the first shape is *closed*. Demonstrate, using chalk, that if you are inside you cannot get out without crossing a line.

3. Point to where the second shape is *open*. Then use a piece of chalk to show that you can go anywhere without crossing a line.

4. Draw several open and closed shapes on the board. As you point to each one, have students identify whether or not the shape is open or closed.

5. Check students' comprehension of these words by asking them to draw open and closed shapes.

corner

1. Show students a rectangle or other geometric shape and identify each *corner* in it.

2. Ask a student to walk to a corner of the floor.

command

1. Explain that a *command* tells you something to do, such as to walk to the corner of the floor.

2. Give several commands for students to follow, such as stand up, sit down, raise your hand.

3. Let students give commands to one another.

Contents

Overview

The units in *Investigations in Number, Data, and Space®* ask teachers to think in new ways about mathematics and how students best learn math. Units such as *Turtle Paths* add another challenge for teachers—to think about how computers might support and enhance mathematical learning. Before you can think about how computers might support learning in your classroom, you need to know what the computer component is, how it works, and how it is designed to be used in the unit. This Tutorial is included to help you learn these things.

The Tutorial is written for you as an adult learner, as a mathematical explorer, as an educational researcher, as a curriculum designer, and finally —putting all these together—as a classroom teacher. Although it includes parallel (and in some cases the same) investigations as the unit, it is not intended as a walk-through of the student activities in the unit. Rather, it is meant to provide experience using the computer program *Geo-Logo*™ and to familiarize you with some of the mathematical thinking in the unit.

The first part of the Tutorial is organized in sessions parallel to the unit. Included in each session are detailed step-by-step instructions for how to use the computer and the *Geo-Logo* program, along with suggestions for exploring more deeply. Many sessions include suggested questions for reflecting on your mathematical thinking. The second part of the Tutorial includes more detail about each component of *Geo-Logo* and can be used for reference while working through the Tutorial or later during the unit. There is also detailed help available in the *Geo-Logo* program itself.

In *Turtle Paths*, students use *Geo-Logo*, a learning environment designed for mathematical, particularly geometric, exploration. Using *Geo-Logo*, students are able to construct paths and geometric shapes in addition to observing them. Since one of the best ways to learn something is to teach it, *Geo-Logo* uses the metaphor of "teaching the turtle" how to move, turn, and draw. Writing a list of instructions for how to construct a shape encourages students to think carefully about geometric properties and to use geometry-oriented language.

Geo-Logo is a rich learning environment intended to be used for open exploration. In the unit, students are first introduced to the environment by playing two games. The later activities are progressively more open, ending with "Free Explore." The Tutorial is organized in the same way, beginning with directed tasks intended to help you become familiar with the environment and commands, and then opening up for you to explore more on your own. For this reason, it might be best to start at the beginning and work through the sessions in order. Teachers new to using computers and *Geo-Logo* can follow the detailed step-by-step instructions. Teachers with more experience might follow the main directions without needing to read all the step-by-step instructions.

As is true with learning any new approach or tool, you will make mistakes, be temporarily stumped, go down wrong paths, test out hypotheses, and so on. This is all part of learning but may be doubly frustrating because you are dealing with computers. It might be helpful to work through the Tutorial and the unit in parallel with another teacher. If you get particularly frustrated, ask for help from the school computer coordinator or another teacher more familiar with using computers. It is not necessary to complete all the sessions in the Tutorial before beginning to teach the unit. You can work through the sessions in parts as you prepare for parallel investigations in the unit.

Although the Tutorial will help prepare you for teaching the unit, you will learn most about *Geo-Logo* and how it supports the unit as you work side-by-side with your students.

Note to teachers: These directions assume *Geo-Logo* Turtle Paths has been installed on the hard disk of your computer. If not, see How to Install *Geo-Logo* on Your Computer, p. 128.

☞ 1. **Turn on** the **computer** following the usual procedure for your computer or by doing the following:

    a. If you are using an electrical power surge protector, switch to the **ON** position.

    b. Switch the **computer** (and the monitor, if separate) to the **ON** position.

    c. Wait until the desktop appears.

Your computer screen may look something like this:

☞ 2. **Open** *Geo-Logo* by doing the following:

    a. **Double-click** on the *Geo-Logo* Turtle Paths folder icon if it is not already open. To double-click, click twice in rapid succession without moving the pointer.

    b. **Double-click** on the *Geo-Logo* Turtle Paths **icon** in this folder.

    c. **Wait** until the *Geo-Logo* opening screen appears. Single click anywhere on the window.

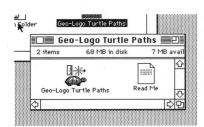

Depending on your computer, you may see other things on the screen in addition to or behind the *Geo-Logo* windows. These are other computer functions that may be available to you but are not part of *Geo-Logo*. If you click one of these by mistake, you can return to *Geo-Logo* by clicking into any *Geo-Logo* window or selecting it from the desktop. For additional information, see Trouble-Shooting, p. 122.

Start an **activity** or game by doing the following:

☞ 1. **Click** on the **[Get the Toys]** button (or whichever activity you want to start).

When you choose an activity, the Tool window, Command Center, Drawing window, and Teach window for that activity fill the screen.

A dialogue box appears with directions for how to get started.

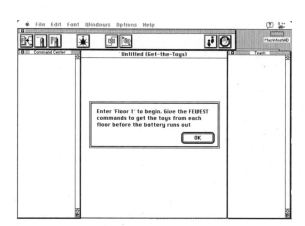

☞ 2. **Click** on **[OK]** or press the **<return>** key to close the dialogue box.

The Get the Toys game is on the computer screen.

Should you need them, trouble-shooting notes are included on p. 122.

How to Start an Activity

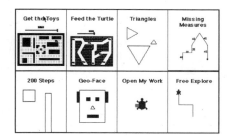

Assistance with windows, vocabulary (commands), tools, directions, and hints is available from the **Help** menu at any time.

☞ 1. Instruct the program to draw the first-floor plan by doing the following:

    a. **Type** floor 1 in the Command Center. (The blinking vertical line, called the text cursor, shows where any typed text will go. It is currently in the Command Center.)

    b. **Press** the **<return>** key.

The game already contains the commands that draw floor plans in a procedure called floor. When you type floor 1 in the Command Center and press **<return>**, the program carries out this group of commands, called a procedure, and draws the map of floor 1.

How to Play Get the Toys on Floor 1

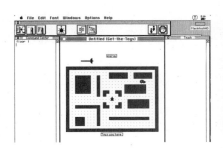

The object of Get the Toys is to write commands for the turtle to follow to get from where it is starting at the elevator to the toy car and back again. The turtle robot knows commands to move and to turn, such as—

fd 50	forward 50	moves the turtle forward 50 steps (or whatever number you want)
bk 10	back 10	moves the turtle back 10 steps (or whatever number you want)
rt 90	right turn 90 degrees	turns the turtle right 90 degrees (use only multiples of 90 for Get the Toys)
lt 90	left turn 90 degrees	turns the turtle left 90 degrees (use only multiples of 90 for Get the Toys)

The turtle starts from the elevator in the middle of the floor.

☛ 2. **Enter commands** in the Command Center to make the turtle go from the elevator to the car. Try your own commands or follow these steps:

a. **Type** fd 60 (f d **<space>** 6 0).

You can use the **<delete>** key to make changes, if needed.

b. **Press <return>**.

Notice that each dot on the floor plan represents 10 turtle steps.

c. **Type** rt 90.

d. **Press <return>**.

Notice that the turtle rotates 90° on its belly to make a right turn. Turn rays are displayed to show the rotation.

e. **Continue** to type commands to make the turtle go to the car.

Notice that when the turtle reaches the toy car, the car moves to the bottom of the screen to an area labeled "Toys you have:" and a dialogue box appears.

f. **Read** the directions in the dialogue box and **click** on **[OK]**.

Got it! Return to the elevator. Think of the commands you used to get here.

OK

The turtle is a robot that follows certain *Geo-Logo* commands. If it does not understand a command, it will write a message in a dialogue box. To get help with what a message means, see *Geo-Logo* Messages, p. 125.

To edit (change) your commands, use the **<delete>** key to back over and erase errors. Type new text and press **<return>**. Another way to edit commands is to use the mouse to select words or blocks of text by dragging (pressing and holding down the mouse button as you move the mouse) over the text. Then press **<delete>** and type new text.

Each time you change a command and press **<return>**, that part is redrawn using the new command.

A third way to change commands is by using *Geo-Logo* tools. Above the Command Center is a row of icons for tools that are available for this activity. For example, to use the tools to erase commands, click

Erase One 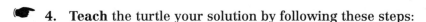 (erases just one command)

Erase All (erases everything in the Command Center).

Each turtle move uses energy from the battery. The battery for each floor is limited to encourage you to use the fewest and most efficient commands. If you go back and delete or combine commands, you will regain some energy.

☞ 3. **Enter commands** to make the turtle go back to the elevator in the middle of the floor. On your return trip you can retrace steps. Try different paths. The turtle does not have to turn around. Remember to press **<return>** after each command you type.

A dialogue box appears when you get back to the elevator after you reach the toy car, giving directions for the next part and telling how many moves in this part.

Click [**OK**] to close the dialogue box.

You have now completed floor 1. Before you finish using the computer or start on floor 2, complete Steps 4 and 5. Then save your game.

☞ 4. **Teach** the turtle your solution by following these steps:

a. **Click** on the **Teach tool** 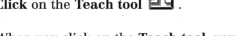 .

When you click on the **Teach tool**, you define the list of commands in the Command Center as a procedure. The computer will ask you to type a name for the procedure.

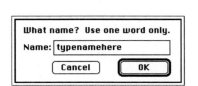

b. **Type** a one-word name for this procedure such as **car1**. You can use letters and numbers. (Since "floor" has already been used to name a procedure, you can not use that name again.)

c. **Click [OK]** or press the **<return>** key.

The procedure appears in the Teach window. It is defined by the name you give it. Notice that the computer adds a first and last line to your commands when you define them as a procedure. The first line says to *car1* (or the name you chose) and the last line says *end*.

Notice that your procedure also includes the procedure floor that draws the floor plan.

Notice that the Drawing window clears in preparation for your next entry.

You have taught the turtle how to move along a path in a certain way. If you want the turtle to do this again, run your procedure following Step 5.

☞ 5. **Run** your **procedure** by following these steps:

a. **Enter** your **procedure** by typing its name, for example car 1, in the Command Center.

If needed, move the text cursor (the blinking vertical line that shows where any typed text will go) into the Command Center by clicking the mouse in that window.

b. **Press <return>**.

If you want to make changes to your procedure, you may edit any commands in the Teach window. Click your cursor in that window, move up or down with the mouse or arrow keys, use the **<delete>** key to erase, and type your changes.

☞ 6. **Choose** the **Help** menu and explore what assistance is available by choosing **Windows**, **Vocabulary** (Commands), **Tools**, **Directions**, and **Hints**.

☞ 7. **Click** on the **Erase All tool** 🗑 to clear the Command Center and Drawing window in preparation for starting floor 2.

Notice that any procedures you have defined are still available in the Teach window.

Before you continue to floor 2 and floor 3 (p. 99) or finish (p. 98), save your game by following the directions on the next page.

When you turn off the computer or start a new activity, the computer memory is cleared of all commands and procedures to make room for new ones. To avoid losing your work, you can save it on a disk before the computer memory is cleared. Once your work has been saved on a disk, you can open it again to show it to someone or to work on it some more.

How to Save Your Game or Work

☞ 1. Choose **Save My Work** from the **File** menu:

 a. **Move** the **mouse pointer** over the word **File** in the menu title bar along the top of the screen.

 b. **Press** and **hold** the mouse **button** until the menu items appear.

 c. Continue to press the mouse button as you **drag** the **pointer** down and select **Save My Work**.

 d. **Release** the mouse **button**.

 (⌘S—To save using the keyboard, hold down the **<⌘>** key and press the **<S>** key.)

 The first time you save your game, a dialogue box, such as the following, will appear asking for a name.

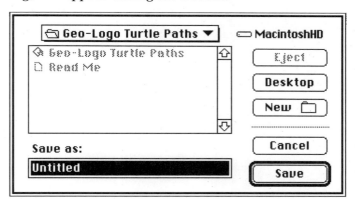

☞ 2. Press **<delete>** to erase **Untitled**.

☞ 3. **Type** a name for your game such as **GW Toys 9/23**. (You can choose any name for a program. However, useful program names include information that helps you find them again, such as the person, activity, and date. Spaces can be used in the titles of saved work.)

☞ 4. **Click** on **[Save]**.

 Notice that the name of your game now appears in the title bar of the Drawing window.

When you save your work this way, a copy is stored on the computer disk.

You can stop using the computer for a while or come back to this game again at another time.

At this time, you might be ready to continue your game on floor 2. If so, skip the next section and come back to How to Finish an Activity when you want to stop playing.

How to Finish an Activity

☛ 1. **To finish** working on this **activity**:

Choose **Close My Work** from the **File** menu. A dialogue box may appear asking whether you wish to save your work, if you have not already saved it or if you have made any changes.

Notice that the computer is ready to start new work on this activity. If others will be playing the game, you may want to leave the screen like this instead of quitting *Geo-Logo* and shutting down the computer.

☛ 2. **To finish** using *Geo-Logo*:

Choose **Quit** from the **File** menu.

☛ 3. **To finish** using the **computer**:

a. Follow the usual procedure to shut down your computer.

b. **Turn off** the **computer** (and monitor, if separate).

If needed, start your computer, open *Geo-Logo*, and select **[Get the Toys]** (or whatever activity you wish) using steps explained earlier.

☞ 1. **To continue** with your previous **game:**

    a. **Choose Open My Work** from the **File** menu.

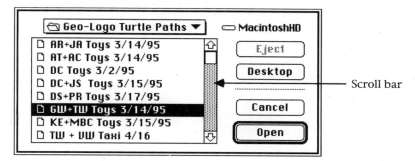

Scroll bar

    If your game has been saved on a different disk, insert that disk, click **[Desktop]**, and choose that disk from the menu.

    b. **Select** a game by clicking on its **title** in the list. You may need to scroll up and down the list to find your title. Click on the up or down arrows in the scroll bar.

    c. **Click** on **[Open]** button (or double-click on the title)

How to Continue with a Saved Game

To play Get the Toys on floor 2 or floor 3, you may begin with a saved game in which you have completed and saved each previous floor. See the section above, How to Continue with a Saved Game.

☞ 1. If needed, **clear** the Command Center and the Drawing window by clicking on the **Erase All tool** 🖼️ .

Notice that all available procedures remain in the Teach window.

☞ 2. **Enter** the command for **floor 2** or **floor 3** plan:

    Type floor 2 in the Command Center and **press** the **<return>** key.

☞ 3. **Enter commands** to make the turtle go from the elevator to the train or rabbit. Note that for Get the Toys, turns must be 90°—for example, rt 90 or lt 90. Moves can be any length—for example, fd 25 or bk 120, although things work out more smoothly, as you might learn, if they are all multiples of 10.

How to Play Get the Toys on Floor 2 and Floor 3

There are a number of different routes the turtle could take, using different numbers of moves and various amounts of battery energy. You might try two or three different paths and compare the numbers of moves by entering commands in the computer for each. Or you might prefer to think through the paths first before entering the commands.

Continue entering commands until the turtle reaches the train or rabbit. It might take a few tries to get to where you want to go. You can edit (change) anything you have already typed by using the mouse or arrow keys to move to that text and the **<delete>** key. When you retype the new command or number, the turtle will erase the old move or turn and follow your new command.

☞ 4. **Enter commands** to make the turtle return to the elevator.

☞ 5. **Teach** the turtle this procedure by clicking on the **Teach tool** and choosing a name (different from your first procedure name).

☞ 6. **Run** your procedure from the Command Center and make any desired changes.

☞ 7. **Click on Erase All tool** .

☞ 8. **Save** your current program by doing the following:

Select **Save My Work** from the **File** menu or ⌘S from the keyboard by holding down the **<⌘>** key and pressing the **<S>** key.

All the new procedures you created in this section are saved along with the procedures you already saved.

Ready for a new challenge? Retrieve the helicopter on floor 9! (Type floor 9 to get there.)

More About Get the Toys

Remember that assistance is available from the **Help** menu at any time: Choose **Windows, Vocabulary** (Commands), **Tools, Directions,** or **Hints.**

Special tools are available in each activity. To use one of these tools, click on its icon in the Tool window.

The Tool window for Get the Toys:

Teach — Teaches the turtle the commands in the Command Center as a new procedure.

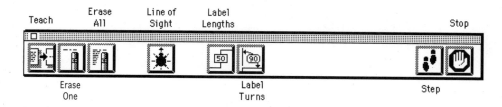

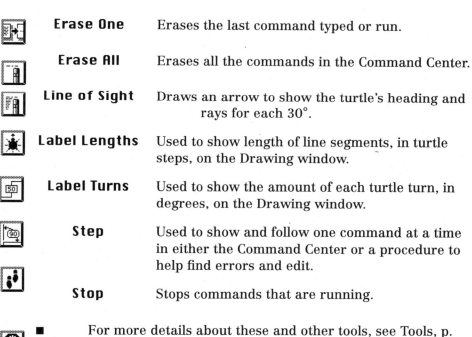

Erase One — Erases the last command typed or run.

Erase All — Erases all the commands in the Command Center.

Line of Sight — Draws an arrow to show the turtle's heading and rays for each 30°.

Label Lengths — Used to show length of line segments, in turtle steps, on the Drawing window.

Label Turns — Used to show the amount of each turtle turn, in degrees, on the Drawing window.

Step — Used to show and follow one command at a time in either the Command Center or a procedure to help find errors and edit.

Stop — Stops commands that are running.

■ For more details about these and other tools, see Tools, p. 120.

- Choose **Show Notes** from the **Windows** menu to record thoughts and observations about your work. (See example below.) To close Notes, choose **Hide Notes** from the **Windows** menu or click in the close box on the left of the Notes title bar.

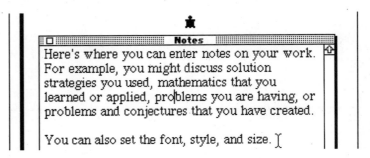

- You can print an entire game, including the picture and list of commands and procedures. Choose **Print** from the **File** menu.

- You can print a single window if you want only a list of your commands, or a copy of your picture or notes. Click into that window and choose **Print Window** from the **File** menu.

- If the drawing window does not print completely, select "Color/Grayscale" in the Print dialog box.

- You can enlarge all the letters in the Command Center and Teach window for easier viewing. Select **All Large** from the **Font** menu. Select **All Small** to change back to the normal font size.

Feed the Turtle

To choose a new activity:

☞ 1. Choose **Change Activity** from the **File** menu.

☞ 2. **Single-Click** on a new activity button on the activity screen.

Get the Toys	Feed the Turtle	Triangles	Missing Measures
200 Steps	Geo-Face	Open My Work	Free Explore

Open *Geo-Logo*, and select **[Feed the Turtle]** following Steps 1–2 in Getting Started with *Geo-Logo* on p. 91.

A dialogue box appears with directions.

> Enter 'pond' to begin. Give the FEWEST commands to get to each piece of food before the turtle's energy runs out.
>
> [OK]

Click **[OK]**.

☞ 1. Type pond and press **<return>** to instruct the program to draw the pond.

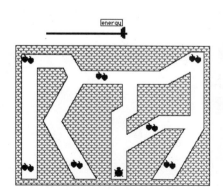

The object of Feed the Turtle is to write commands that the turtle follows to move along the paths in the pond to get to the berries. Each time the turtle reaches food, it gets more energy.

In Feed the Turtle, the turtle knows these move and turn commands:

fd and bk   Only use multiples of 10—for example, fd 40 or bk 130.

rt and lt   Only use multiples of 30—for example, rt 60 or lt 120.

☞ 2. **Enter commands** to move the turtle along a path to get to the berries.

Notice that you will have to use turn commands other than 90°.

You might find the **Line of Sight tool** helpful for estimating turns. Press and hold down the tool button to display turn rays.

☞ 3. **Continue** to **enter commands** until the turtle has moved through the pond paths and eaten all the berries.

You might try many different combinations of commands as you move through the paths in the pond. You may want to go back and combine commands to conserve energy. For example, combine fd 40 and bk 10 into fd 30 or lt 120 and rt 30 into lt 90.

Because your command list will get long as you move along the pond paths, you might want to save your game as you are working rather than waiting until the end.

A dialogue box appears when you have reached all the berries and tells how many total moves in your game. Read the comments and click **[OK]**.

☞ 4. **Teach** the turtle your solution by clicking on the **Teach tool**. Type a name for the procedure. (Do not use the word *pond* in the name as it has already been used to name a procedure.) Click **[OK]**.

☞ 5. **Check** your procedure from the Command Center. Make any changes you want in the Teach window.

☞ 6. **Save** your game.

☞ 7. **Print** your game.

    a. Check that your printer is connected properly and turned on.

    b. Choose **Print** from the **File** menu.

☞ 8. **Pause and reflect.** You might want to pause for a bit and reflect on your mathematical thinking as you played these last two games. Here are some questions you might consider:

- What geometric terms and systems of measurement did you use to instruct the turtle to move along various paths?

- What mathematical processes—such as addition, subtraction, estimation, and "undoing"—did you use to play these games?

- How did limited battery energy influence how you thought about paths and how you wrote commands?

Triangles

Open *Geo-Logo* and select [**Triangles**], following Steps 1–2 in Getting Started with *Geo-Logo* on p. 91.

A dialogue box appears with directions.

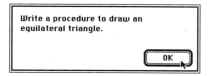

Click [**OK**].

Triangles is not a game but an exploration of how to write turtle commands to draw various equilateral triangles. Unlike the previous two activities, in this exploration your moves and turns are not restricted. The turtle will draw any path you command.

The directions are to write a procedure to draw an equilateral triangle because most students at this level are not prepared for the mathematical thinking involved in writing commands to draw other types of triangles. Equilateral triangles have sides of equal measure and also turns of equal measure. (Students are encouraged to use turns that are multiples of 30).

Remember that the amount the turtle turns to draw an angle may not be the same as the measurement of the angle. In the figure, the turtle turns more than 90°, but the angle measure is less than 90°.

How to Make Triangles

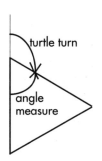

 1. **Enter commands** to make the turtle draw an equilateral triangle. The turtle turn and angle measure always add up to 180°.

   Use the **Erase One tool** to help erase and rewrite as you estimate and try different commands.

   Try using the **Draw Commands tool** to help you estimate turns and distances. See More About Triangles on p. 106 of this tutorial.

   Try using the ht (hide turtle) command if the turtle is covering a part of the triangle you want to see. Use the st (show turtle) command to display the turtle again.

 2. **Edit** your **commands** to make any changes you want.

3. **Teach** the turtle your procedure. **Run** it from the Command Center. **Erase All** to clear the Command Center and Drawing window for the next triangle.

☛ 4. **Explore** the following questions. For each, enter commands, teach the turtle your procedure, run it, and then erase all.

■ Can you draw the biggest possible equilateral triangle that will still fit on the screen?

■ Can you draw the smallest equilateral triangle possible?

■ What would happen if you changed all your right turns to left turns?

■ Can you make a triangle that has a total path length, or perimeter, twice as large as your first one?

■ Try using the `repeat` command as a short cut. For example, `repeat 3 [fd 30 rt 120]` repeats 3 times the `fd` and `rt` commands in the list between the square brackets. In general, it repeats # times [the commands in the list].

■ Can you make an equilateral triangle with a horizontal side?

☛ 5. **Save** your work.

☛ 6. **Pause and reflect.** You might want to pause for a bit and reflect on your mathematical thinking in this activity. You might use the Notes window to record your thinking in this and other activities. Here are some questions you might consider:

■ How could you tell that the smallest triangle was equilateral?

■ What properties of geometric figures did you think about when you wrote *Geo-Logo* commands to draw equilateral triangles?

More About Triangles

The Tool window for Triangles:

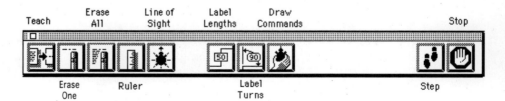

■ The **Ruler tool** measures length, starting at the turtle's position. The other end of the ruler follows a point that you move with the mouse. Click to freeze the ruler and display the length in a dialogue box.

■

This tool	Does this	Resulting in
	Labels the lengths of line segments, in turtle steps.	
	Labels the amount of each turtle turn, in degrees.	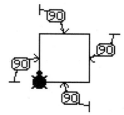

■ The **Draw Commands tool** 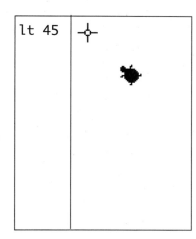 allows you to use the mouse to turn and move the turtle directly as corresponding *Geo-Logo* commands are created automatically and written in the Command Center.

 a. Click on the **Draw Commands tool**.

   Notice that when you move the cursor into the Drawing window, it changes into cross-hairs and the turtle turns to face it. As you move it with the mouse, the turtle continually turns to face in that direction and a corresponding turn command is written and updated in the Command Center.

 b. Move the cross-hairs in the desired direction.

 c. Click to freeze and accept a turn command for the turtle.

   Notice that the cursor now changes into a hand that can be used to move the turtle along a path in the chosen direction.

 d. Drag the turtle to the next location. You can drag it only in the direction in which it is heading.

 e. Release the mouse button to freeze and accept a move command for the turtle.

 f. Continue to create new turn and move commands to complete your drawing.

 g. Click in any other window—for example, the Command Center—to stop using this tool.

■ Choose from the **Options** menu to activate (or deactivate) **Fast Turtle** and **Turn Rays**. (See the **Options** menu on p. 118.)

How to Find Missing Measures

> Write a procedure to draw each figure.
> Find all the missing lengths and turns.
>
> [OK]

☞ 1. **Complete** Student Sheet 15, Missing Lengths and Turns, writing commands to draw each figure.

Open *Geo-Logo*, and select **[Missing Measures]**, following Steps 1–2 in Getting Started with *Geo-Logo* on p. 91.

A dialogue box appears with directions.

Click **[OK]**.

☞ 2. **Enter commands** to make the turtle draw each completed shape. As you try your plans, note any changes you decide to make on the Student Sheet.

Try using the **Turtle Turner tool** ✳ to measure turns on figures such as #6 on Student Sheet 15.

Remember that the **Label Length tool** and the **Label Turn tool** are available.

☞ 3. **Teach** each procedure and **save** your work at the end.

☞ 4. **Pause and reflect.** You might want to pause for a bit and reflect on your mathematical thinking in this activity. Here are some questions you might consider:

- How is your thinking different when you construct shapes using *Geo-Logo* rather than drawing them on paper?

- What mathematical processes—such as quantitative reasoning, mental arithmetic, and logic—did you use to find the missing measures?

More About Missing Measures

The Tool window for Missing Measures:

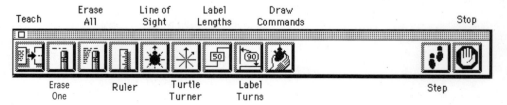

- The **Turtle Turner tool** measures turns from the turtle's heading. Click to select the tool. One arrowhead shows the turtle's heading. The other follows the cursor as it moves with the mouse. Click to freeze the arrowhead and show a turn command. Use this tool to help you estimate measures of turns in figures.

Open *Geo-Logo*, and select **[200 Steps]**, following Steps 1–2 in Getting Started with *Geo-Logo* on p. 91.

A dialogue box appears with directions.

How to Make Rectangles in 200 Steps

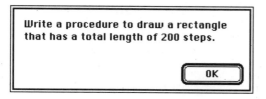

Write a procedure to draw a rectangle
that has a total length of 200 steps.

OK

Click **[OK]**.

☞ 1. **Enter commands** to make the turtle draw a rectangle with a total length of 200 steps.

Try using the ht (hide turtle) command at the end of the procedure to check if the path is closed. Use st (show turtle) to return the turtle to the screen.

Try using the repeat command as a short cut. For more information about the repeat command, see p. 116.

☞ 2. **Enter commands** to draw a number of different rectangles with total lengths of 200 steps. Teach and run each procedure after writing it.

☞ 3. **Enter commands** to draw a rectangle with side lengths that are not multiples of 10. Teach and run your procedure after writing it.

☞ 4. **Enter commands** to draw the paths suggested on Student Sheet 19, 200-Step Paths That Are *Not* Rectangles. Teach and run each procedure.

Try using **Cut** and **Paste** from the **Edit** menu to copy commands you have already written as a short cut when you are writing procedures.

☛ 5. **Pause and reflect.** You might want to pause for a bit and reflect on your mathematical thinking in this activity. Here is a question you might consider:

- Which geometric properties connect numbers with geometric shapes?

More About 200 Steps

The Tool window for 200 Steps:

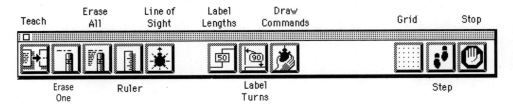

- The **Grid tool** places a dot grid and labeled axes in the Drawing window. Use this tool to estimate distances and to help plan move measurements. Click again to remove. For more details, see The Coordinate Grid, p. 114.

- Try using *Geo-Logo* as a calculator. Enter the command pr 85 + 15. Notice that a print window appears with the answer to the calculation. You can use pr (print command) for many arithmetic calculator functions. Enter ct to clear text in the Print window. Click the close box when finished. Use the **Erase One tool** to erase the pr command line before writing more commands.

Open *Geo-Logo*, and select **[Geo-Face]**, following Steps 1–2 in Getting Started with *Geo-Logo* on p. 91.

A dialogue box appears with directions.

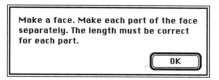

Make a face. Make each part of the face separately. The length must be correct for each part.

OK

Click [**OK**].

The object of the Geo-Face activity is to draw a face with features in the shape of rectangles, squares, or triangles with given perimeters. After designing the face on paper, you will teach the computer each part separately. Then, with the help of a special move-slide-turn feature, you will write commands to assemble the entire face with the features placed where you choose.

☛ 1. On paper: Use Student Sheet 22, Geo-Face Plans, to **design** your Geo-Face and **plan** procedures to draw each part with the given perimeters.

☛ 2. On computer: **Enter commands** for and **teach** the computer how to draw each separate part of the face. Make sure that the turtle ends each procedure facing the same way it started. After you have taught the computer each part, check each procedure by running it. Then erase it before you enter commands for the next procedure.

Notice that a procedure called head already exists to draw the outline.

☛ 3. To **assemble** your face, follow these steps:

a. **Type** head and press **<return>** to draw the outline.

b. **Drag** (press and hold the mouse button as you move the mouse) the turtle to the starting location for each part. Notice that this adds two commands in the Command Center—make-points and jumpto. (For more information about these commands, see p. 114, More About *Geo-Logo*.)

c. **Enter** the procedure name for the part and press **<return>**.

d. **Continue** to move the turtle and enter procedure names until you have placed all the parts.

Try using the move-slide-turn feature to position the turtle or to slide or turn the face parts. (See Examples of Motions, p. 119.)

How to Make a Geo-Face

If your computer has color capability, try making the parts of your face different colors. Add a `setc` command (for example, `setc red`) at the beginning of the procedure that draws the part. For a list of colors, see `setc` command, p. 116.

☞ 4. When your Geo-Face is complete, **teach** the turtle this group of commands and procedures as a new procedure. Enter one word to have the turtle draw the entire face with all the parts.

☞ 5. **Save** your work at the end.

☞ 6. **Print** your work.

☞ 7. **Pause and reflect.** You might want to pause for a bit and reflect on your mathematical thinking in this activity. Here are some questions you might consider:

■ What problem-solving strategies, such as looking for patterns, trial and error, or drawing a picture on paper first, did you use?

■ How is your thinking different observing and drawing geometric shapes off-computer as compared to on-computer using *Geo-Logo*?

More About Geo-Face

The Tool window for Geo-Face:

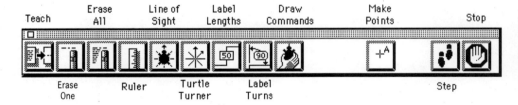

■ The **Make Points tool** and the `jumpto` command can be used together to move the turtle to any desired location. Click the tool. Click to place points in the Drawing window. These points are automatically labeled with letters. Write a `jumpto` command that can be used to move the turtle to a point without leaving a path. For example, `jumpto A` moves the turtle to point A placed with the make-points tool or command.

■ The Motions feature is available in Geo-Face. See Motions, p. 119.

Free Explore

The Free Explore activity is available for you to use to extend and enhance activities in the unit and as an environment to further explore *Geo-Logo*.

In Free Explore, the turtle responds to all *Geo-Logo* commands, and all tools are available from the Tool window.

More About *Geo-Logo*

Help

Assistance is available as you work with *Geo-Logo* activities. From the **Help** menu, choose any of the following:

Windows provides information on *Geo-Logo*'s three main windows: Command Center, Drawing, and Teach.

Vocabulary provides a listing of *Geo-Logo*'s commands and examples.

Tools provides information on *Geo-Logo*'s tools (represented on the Tools window as icons).

Directions provides instructions for the present activity.

Hints gives a series of hints on the present activity, one at a time. It is dimmed when there are no available hints.

The Coordinate Grid

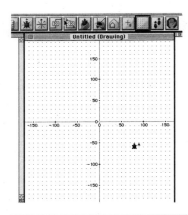

The turtle's location in the Drawing window can be described as a position on a coordinate grid with the origin [0 0] in the middle of the screen where the turtle starts. The tool to display this grid is available in 200 Steps and Free Explore.

In *Geo-Logo*, locations on this grid are described using two numbers, without a comma, within square brackets. For example, the turtle in this picture is at [80 –60]. Commands such as jumpto [80 –60] use coordinates for locations.

For those familiar with the coordinate system of locating a point on a plane, displaying the coordinate grid can be useful for writing turtle movement commands in *Geo-Logo*. The coordinate system and this tool are explored in detail in the grade 4 unit *Sunken Ships and Grid Patterns*.

Commands

The following commands are available in *Geo-Logo*. Some have been previously introduced in the investigations where they are most useful.

Command	What It Means	What It Does
bk 10	back 10	Moves the turtle back 10 steps (use whatever number you wish). The turtle leaves a path if its pen is down. *See also* pd *and* pu.
ct	clear text	Clears, or erases, all the text in the Print window.

eraseall	erase all	Erases all the commands in the Command Center.
fd 50	forward 50	Moves the turtle forward 50 steps (use whatever number you wish). *See also* pd *and* pu.
fill	fill	Fills a closed shape or the entire Drawing window with the current turtle's color, starting at the current turtle's position. If the turtle's pen is over a path, only that path is filled. To fill a shape, use pu, then fd and rt or lt to move inside the shape. Use setc to set the color, then fill. **Note:** If the shape you want to fill is on a grid, turn off the grid first before filling. Use the hp command to hide points if you want to use jumpto or motions with fill.
hp	hide points	Hides points (they become invisible).
ht	hide turtle	Hides the current turtle (it becomes invisible).
jumpto [__ __]		Moves the turtle to the point whose coordinates are in the brackets without leaving a path. Can use jt as abbreviation for jumpto.
jumpto A		Moves the turtle to point A without leaving a path. The letter must first be defined as a point.
lt 120	left 120	Turns the turtle left 120 (or any number of) degrees.
make-points [A [20 50] B [-70 -100]]		Makes points and shows them on the Drawing window. See the **Make Points tool**, which automatically generates this command.
pd	pen down	Puts the turtle's pen down so that when it moves, it draws a path.
print [My drawing]		Prints whatever is within brackets or acts as a calculator. For example, if you type pr 85 + 15, *Geo-Logo* will print 100.

print colors		Prints a list of colors available to use with the setc command.
pu	pen up	Puts the turtle's pen up so that when it moves, it does not draw a path.
repeat 4 [fd 10 rt 90]		Repeats the commands in the list the specified number of times; in this example, 4 times. (The list is whatever is between the [square brackets].)
rt 45	right 45	Turns the turtle right 45 degrees (or any number of degrees).
setc black	set color	Sets the turtle's color; this affects the color of the turtle and the color for drawing and filling. The color names are: white black gray gray2 gray3 yellow orange red pink violet blue blue2 green green2 brown brown2.
sp	show points	Shows points and labels.
st	show turtle	Shows the turtle.

Menus

How to Use Menus

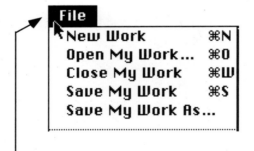

Point to the menu you want and press the mouse button...

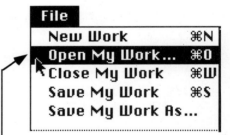

...then drag the selection bar to your choice and release the button

Some menu choices are also available from the keyboard. On the menu, the ⌘N indicates that, instead of selecting the choice from the menu, you could type ⌘-N. Hold down the Command key (with the ⌘ and symbols on it) then press the **<N>** key.

A menu choice may be dimmed indicating it is not available in a particular situation.

Geo-Logo's Menus

The **File** menu deals with documents and quitting.

New Work starts a new document.

Open My Work opens previously saved work.

Close My Work closes present work.

Save My Work saves the work.

Save My Work As saves the work with a new name or to a different disk or folder.

Print prints a whole document.

Print Window prints only the active window (the last one clicked on).

Quit quits *Geo-Logo*.

The **Edit** menu contains choices to use when editing your work:

Undo reverses the last thing done, such as **Delete**, **Cut**, **Erase**, and **Erase All**.

Cut deletes the selected object and saves it to a space called the clipboard.

Copy copies selected object to the clipboard.

Paste places the contents of the clipboard in the cursor location.

Clear deletes the selected object (works the same as using the **<delete>** key).

Stopall stops running the procedure.

The **Font** menu is used to change the appearance of text. The change applies to the active window (the Command Center, Teach, Print, or Notes windows).

The first names are choices of typeface.

Size and **style** have additional choices; pull down to select them and then to the right. See the example for **Style** shown at right. The **Size** choice works the same way.

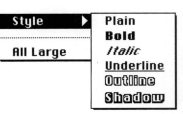

All Large changes all text in all windows to a large-size font. This is useful for demonstrations. This selection toggles (changes back and forth) between **All Large** and **All Small**.

Windows

Hide Command Center
Hide Drawing
Hide Teach
Hide Tools

Show Print
Show Notes

Options

Fast Turtle
✓ Turn Rays

Decimal Places ▶

Help

Windows...
Vocabulary...
Tools...
Directions...   ⌘D
Hints...   ⌘H

The **Windows** menu shows or hides the windows. If you hide a window, such as the Drawing window, the menu item changes to **Show** followed by the name of the window—for example, **Show Drawing**. You can also hide a window by clicking in the "close box" in the upper-left corner of the window.

The **Show Print** choice opens the Print window and displays text generated by a print command from the Command Center. The **Notes** choice opens and hides the Notes window. You can use this window to enter and keep permanent notes.

The **Options** menu allows you to customize *Geo-Logo*.

Fast Turtle turns the turtle quickly and so speeds up drawing. Usually in *Geo-Logo*, the turtle turns slowly, to help students build images of the turns.

Turn Rays displays rays during turns to help visualize the turn. After a turn command is entered, a ray is drawn to show the turtle's initial heading. Then as the turtle turns, another ray turns with it, showing the change in heading throughout the turn. A ray also marks every 30° of turn.

Decimal Places controls how many numbers after the decimal point (i.e., 10ths, 100ths...) are printed by certain commands and tools, such as the **Ruler, Turtle Turner, Label Lines,** and **Label Turns tools.** If 0 (zero) numbers are shown, the number is rounded to the nearest integer.

The **Help** menu provides assistance.

Windows provides information on *Geo-Logo*'s three main windows: Command Center, Drawing, and Teach.

Vocabulary provides a list of *Geo-Logo*'s commands and examples.

Tools provides information on *Geo-Logo*'s tools (represented on the Tools window as icons).

Directions provides instructions for the present activity.

Hints gives a series of hints on the present activity, one at a time. It is dimmed when there are no available hints.

Examples of Motions

The following motions are automatically available in Geo-Face and Free Explore.

Move turtle To move the turtle, click on the turtle and drag it by pressing and holding the mouse button while you move the mouse. Two commands, make-points and jumpto, are automatically placed in the Command Center.

Click...   ...then drag to move the turtle.

Move point Click on the point (the +) and drag it to a new location.

Click...   ...then drag to move a point.

If the point is hidden under a turtle, click on the label (for example, the letter *C*) and the point will jump to the cursor to be moved.

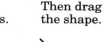

You can slide and turn any procedure that has been defined in the Teach window.

Slide To slide a procedure, click in the middle of one of its *line segments* and drag it to a new location. Release the mouse button when the proce-dure is in the correct location.

Click here...   ...and this cursor appears.   Then drag the shape.

A slide command is automatically placed in the Command Center.

Turn Press the **<shift>** key, click on a corner or line segment, and drag in a circle to turn the proce-dure. Release the mouse button when the procedure has the desired orientation.

Click here...   ...and this cursor appears.   Then drag to turn the shape.

A turnit command is automatically placed in the Command Center which can be edited to make fine adjustments.

Note: You cannot drag the corner at which the turtle starts drawing the pro-cedure because it is that point around which the procedure turns.

Tools

Only the most commonly used tools are available and displayed for each activity. All tools are available for Free Explore.

Click on a tool to use it.

Teach

Teaches the turtle your procedure. Give the procedure a name. Enter its name in the Command Center to run it.

Ruler

Move the cursor to a point on the Drawing window. Click to find the length from the turtle to that point.

Line of Sight

Click and hold the mouse button to see a turtle turner show the turtle's heading.

Turtle Turner

Move the cursor on the Drawing window to turn the arrow. Click to show the turn command.

Draw Commands

Move the pointer and click to create a turn command. Then Grab the turtle and pull it forward or back.

Draw Commands

Click on a line segment or corner and drag it to its new location.

Grid

Puts a grid on the Drawing window, or removes it.

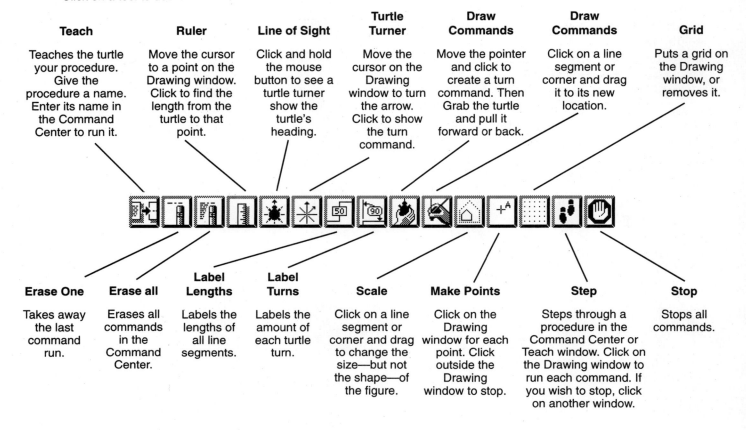

Erase One

Takes away the last command run.

Erase all

Erases all commands in the Command Center.

Label Lengths

Labels the lengths of all line segments.

Label Turns

Labels the amount of each turtle turn.

Scale

Click on a line segment or corner and drag to change the size—but not the shape—of the figure.

Make Points

Click on the Drawing window for each point. Click outside the Drawing window to stop.

Step

Steps through a procedure in the Command Center or Teach window. Click on the Drawing window to run each command. If you wish to stop, click on another window.

Stop

Stops all commands.

The *Geo-Logo* screen looks like this:

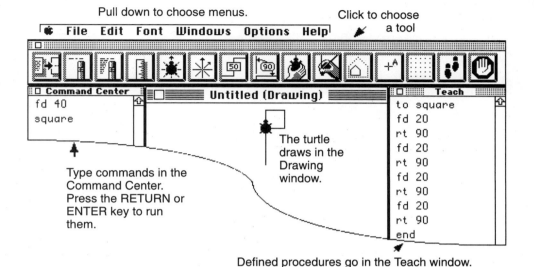

Pull down to choose menus.

Click to choose a tool

Type commands in the Command Center. Press the RETURN or ENTER key to run them.

The turtle draws in the Drawing window.

Defined procedures go in the Teach window.

Command Center

Type commands you wish the turtle to run immediately in the Command Center. Press the **<return>** key after each command. Make changes to commands directly in the Command Center; they are reflected automatically in the drawing when you press **<return>**.

Teach Window

When you have a sequence of commands you might wish to use again, you can define them as a procedure. Click the **Teach tool** . A dialogue box appears, asking for a one-word name for the procedure. The name (with the word *to* in front of it) and the commands (with *end* added) are placed in the Teach window on the right and named as a defined, or taught, procedure. You can then type the name of the procedure as a new command.

If you change a procedure in the Teach window (for example, changing each fd 20 to fd 30 in the procedure square above), the change will be reflected in the Drawing window as soon as you click out of the Teach window.

In addition, the **Draw Commands tool** and the **Change Shape** tool, described in Tools on p. 120, allow you to change the geometric figure directly and see the effect on the *symbols* reflected immediately in the Command Center.

This section contains suggestions for how to correct errors, how to get back to what you want to be doing when you are somewhere else in the program, and what to do in some troubling situations.

If you are new to using the computer, you might also ask a computer coordinator or an experienced friend for help.

No *Geo-Logo* Icon to Open

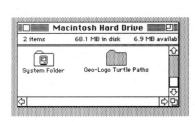

■ Check that *Geo-Logo* Turtle Paths has been installed on your computer by looking at a listing of the hard disk.

■ Open the folder labeled *Geo-Logo* Turtle Paths by double-clicking on it.

■ Find the icon for the *Geo-Logo* Turtle Paths application and double-click on it.

Geo-Logo Turtle Paths

Nothing Happened After Double-Clicking on the *Geo-Logo* Icon

■ If you are sure you double-clicked correctly, wait a bit longer. *Geo-Logo* takes a while to open or load and nothing new will appear on the screen for a few seconds.

■ On the other hand, you may have double-clicked too slowly or moved the mouse between your clicks. In that case, try again.

In Wrong Activity

■ Choose **Change Activity** from the **File** menu.

Text Written in Wrong Area

■ Delete text.

■ Click cursor in the desired area or on the desired line and retype text (or select text and use **Cut** and **Paste** from the **Edit** menu to move text to desired area).

Out of Room in Command Center

■ Continue to enter commands. Text will scroll up and old commands will still be there but temporarily out of view. To scroll, click on the up or down arrows in the scroll bar along the right side of the window.

A Window Closed by Mistake

■ Choose **Show Window** from the **Windows** menu.

Windows or Tools Dragged to a Different Position by Mistake

■ Drag the window back into place by following these steps: Place the pointer arrow in the stripes of the title bar. Press and hold the button as you move the mouse. An outline of the window indicates the new location. Release the button and the window moves to that location.

I Clicked Somewhere and Now *Geo-Logo* Is Gone! What Happened?

You probably clicked in a part of the screen not used by *Geo-Logo*, and the computer therefore took you to another application, such as the "desktop."

■ Click on a *Geo-Logo* window, if visible.

■ Double-click on the *Geo-Logo* Turtle Paths program icon.

The Turtle Disappeared off the Screen. Why?

■ If a command moves the turtle off the screen, write the opposite command to make it return. For example, if fd 500 sent the turtle off the screen, bk 500 will return it.

   Note: Many versions of Logo "wrap"—that is, when the turtle is sent off the top of the screen, it reappears from the bottom. *Geo-Logo* does not wrap when it is opened because students are learning to connect *Geo-Logo* commands to the geometric figures they draw.

How Do I Select a Section of Text?

In certain situations, you may wish to copy or delete a section, or block, of text.

■ Point and click at one end of the text. Drag the mouse by holding down the mouse button as you move to the other end of the text. Release the mouse button. Then use the **Edit** menu to **Copy**, **Cut**, and **Paste**.

System Error Message

Some difficulty with the *Geo-Logo* program or your computer caused the computer to stop functioning.

- Turn off the computer and repeat the steps to turn it on and start *Geo-Logo* again. Any work you saved will still be available to open from your disk.

I Tried to Print and Nothing Happened

- Check that the printer is connected and turned on and that "Color/Grayscale" is selected in the Print dialog box.

- When printers are not functioning properly, a system error may occur causing the computer to "freeze." If there is no response from the keyboard or when moving or clicking with the mouse, you may have to shut down the computer and start over.

Geo-Logo Messages

The turtle responds to *Geo-Logo* commands as a robot. If it does not under-stand a command or has a suggestion, a dialogue box may appear with one of the following messages. Read the message, click on **[OK]** or press **<return>** from the keyboard and correct the situation as needed.

Disk or directory full.

> The computer disk is full.
>
> ■ Use **Save My Work As** to choose a different disk.

For rt and lt commands in this activity, use 90.

> In Get the Toys, turn commands only use multiples of 90.
>
> rt 60   needs 90   -> rt 90

For fd and bk commands in this activity, use 10's numbers: 10, 20, 30, 40....

> In Feed the Turtle, move commands only use multiples of 10.
>
> fd 55   needs multiple of 10   -> fd 50 or fd 60

For rt and lt commands in this activity, use 30's numbers: 30, 60, 90, 120, 150, 180....

> In Feed the Turtle, turn commands only use multiples of 30.
>
> rt 65   needs multiple of 30   -> rt 60

I don't know how to *name*.

> Program does not recognize the command as written.
>
> fd50   needs a space between fd and 50 -> fd 50
>
> fdd 50   extra "d"
>
> mypicturje   misspelling

I don't know what to do with *name*.

> You may have given too many inputs to a command or no command at all.

> > `fd 50 30`   needs only one number

> You may have left out a command.

> > `5 + 16`   change to `print 5 + 16`

I'm having trouble with the disk or drive.

> The disk might be write-protected, there is no disk in the drive, or some similar problem.

> ■ Use **Save My Work As** to choose a different disk.

***Name* can only be used in a procedure.**

> Certain commands, such as `end` and `stop`, can't be used in the command center.

> ■ Don't use that command if you don't need to.

> ■ Define the procedure in the Teach window.

***Name* does not like *name* as input.**

> A command needs a certain type of input and didn't get it from the command following it.

> > `fd fd 30`   Omit one `fd` or put a number after the first one.

> > `repeat [fd 30 rt 90]`   Repeat needs two inputs; a number and a list—for example, `repeat 4 [fd 30 rt 90]`.

***Name* is already used.**

> A procedure already exists with that name.

> ■ Use a different name.

***Name* needs more inputs.**

> Command `name` needs an input, such as a number.

> > `fd`   needs how much   `-> fd 30`

> > `rt`   needs how much to turn `-> rt 30`

> > `floor`   needs which floor   `-> floor 1 or floor 2`

Number too big.

> There are limits to numbers *Geo-Logo* can use; it can use numbers up to 2147483647.

> ■ Don't exceed the limit.

Ouch!

> Command has caused the turtle to run into a barrier in Get the Toys or Feed the Turtle.

> ■ Rewrite the command.

Out of space.

> There is no free memory left in the computer.

> ■ Enter the command `recycle` to clean up and reorganize available memory.

> ■ Eliminate commands or procedures you don't need.

> ■ Save and start new work.

The maximum *value* for *name* is *number*.

> The input is too high.

> For example, The `maximum value for fd is 9999`.

> ■ Use a smaller number.

The minimum *value* for *name* is *number*.

> The input is too low a number.

> For example, The `minimum value for fd is -9999`.

> ■ Use a higher number.

Turtle is out of energy!

> Too many commands in your procedure in Get the Toys or Feed the Turtle.

> ■ Combine commands in your procedure, then try it again (for example, `fd 30 fd 10 -> fd 40`).

> ■ Find a more direct path.

The *Geo-Logo* disk for *Turtle Paths* that you received with this unit contains the *Geo-Logo* Turtle Paths program and a Read Me file. You may run the program directly from this disk, but it is better to put a copy of the program and the Read Me file on your hard disk and store the original disk for safekeeping. Putting a program on your hard disk is called *installing* it.

Note: *Geo-Logo* runs on a Macintosh II computer or above, with 4 MB of internal memory (RAM) and Apple System Software 7.0 or later. (*Geo-Logo* can run on a Macintosh with less internal memory, but the system software must be configured to use a minimum of memory.)

To install the contents of the *Geo-Logo* Turtle Paths disk on your hard drive, follow the instructions for your type of computer or these steps:

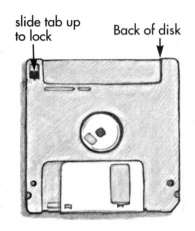

slide tab up
to lock   Back of disk

1. Lock the *Geo-Logo* Turtle Paths program disk by sliding up the black tab on the back, so the hole is open.

   The *Geo-Logo* Turtle Paths disk is your master copy. Locking the disk allows copying while protecting its contents.

2. Insert the *Geo-Logo* Turtle Paths disk into the floppy disk drive.

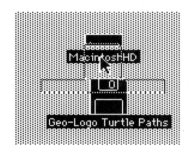

3. Double-click on the icon of the *Geo-Logo* Turtle Paths disk to open it.

4. Double-click on the Read Me file to open and read it for any recent changes in how to install or use *Geo-Logo*. Click in the close box after reading.

5. Click on and drag the *Geo-Logo* Turtle Paths disk icon (the outline moves) to the hard disk icon until the hard disk icon is highlighted, then release the mouse button.

   A message appears indicating that the contents of the *Geo-Logo* Turtle Paths disk are being copied to the hard disk. The copy is in a folder on the hard disk with the name *Geo-Logo* Turtle Paths.

6. Eject the *Geo-Logo* Turtle Paths disk by dragging it to the trash. (Don't worry; the disk will *not* be erased.) Store the disk in a safe place.

7. If the hard disk window is not open on the desktop, open the hard disk by double-clicking on the icon.

   When you open the hard disk, the hard disk window appears, showing you the contents of your hard disk. It might look something like this. Among its contents is the folder labeled *Geo-Logo* Turtle Paths holding the contents of the *Geo-Logo* disk.

☞ 8. Double-click the *Geo-Logo* Turtle Paths folder to select and open it.

When you open the *Geo-Logo* Turtle Paths folder, the window contains the program and a Read Me file.

To select and run *Geo-Logo* Turtle Paths, double-click on the program icon.

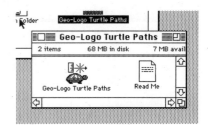

Optional

For ease at startup, you might create an alias for the *Geo-Logo* Turtle Paths program by following these steps:

☞ 1. Select the program icon.

☞ 2. Choose **Make Alias** from the **File** menu.

The alias is connected to the original file that it represents, so when you open an alias, you are actually opening the original file. This alias can be moved to any location on the desktop.

☞ 3. Move the *Geo-Logo* Turtle Paths alias out of the window to the desktop space under the hard disk icon.

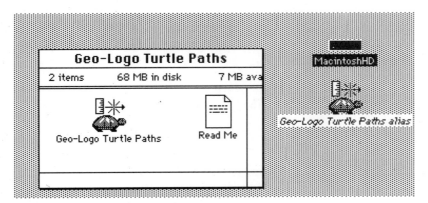

For startup, double-click on the *Geo-Logo* Turtle Paths alias instead of opening the *Geo-Logo* Turtle Paths folder to start the program inside.

Saving Work on a Different Disk

For classroom management purposes, you might want to save student work on a disk other than the program disk or hard disk. Make sure the save-to disk has been initialized (see instructions for your computer system).

☞ 1. Insert the save-to disk into the drive.

☞ 2. Choose **Save My Work As** from the **File** menu.

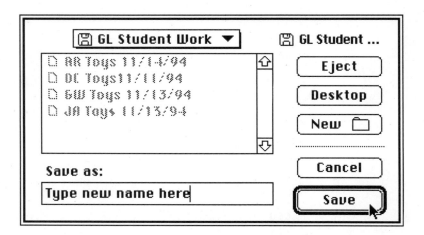

The name of the disk the computer is saving to is displayed in the dialogue box. To choose a different disk, click the **[Desktop]** button and double-click to choose and open a disk from the new menu.

☞ 3. Type a name for your work if you want it to have a new or different name from the one it currently has.

☞ 4. Click on **[Save]**.

Deleting Copies of Student Work

As students no longer need previously saved work, you may want to delete their work (called "files") from a disk. This cannot be accomplished from inside the *Geo-Logo* program. However, you can delete files from disks at any time by following directions for how to "Delete a File" for your computer system.

Blackline Masters

Family Letter   132

Investigation 1

Investigation 2

Investigation 3

General Resources for the Unit

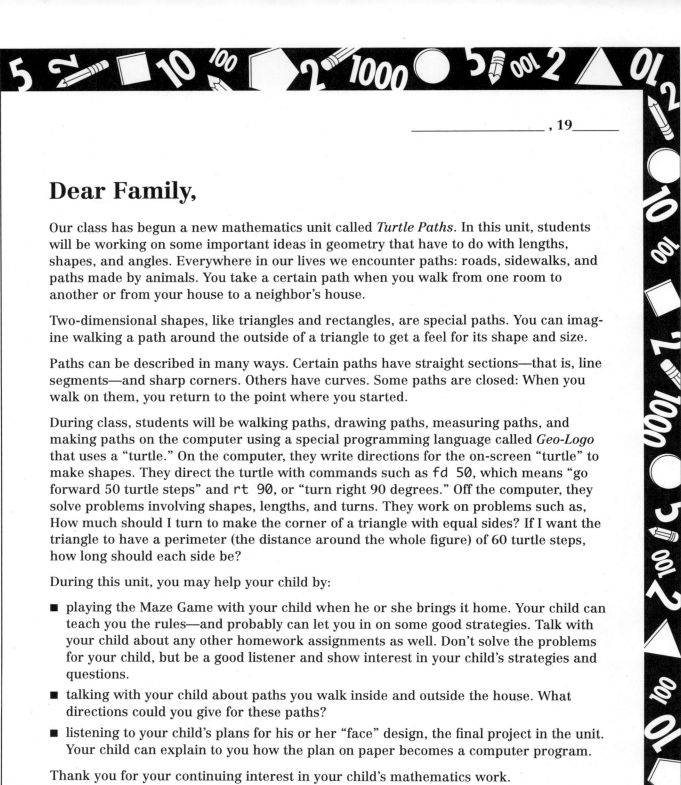

_____ , 19_____

Dear Family,

Our class has begun a new mathematics unit called *Turtle Paths*. In this unit, students will be working on some important ideas in geometry that have to do with lengths, shapes, and angles. Everywhere in our lives we encounter paths: roads, sidewalks, and paths made by animals. You take a certain path when you walk from one room to another or from your house to a neighbor's house.

Two-dimensional shapes, like triangles and rectangles, are special paths. You can imagine walking a path around the outside of a triangle to get a feel for its shape and size.

Paths can be described in many ways. Certain paths have straight sections—that is, line segments—and sharp corners. Others have curves. Some paths are closed: When you walk on them, you return to the point where you started.

During class, students will be walking paths, drawing paths, measuring paths, and making paths on the computer using a special programming language called *Geo-Logo* that uses a "turtle." On the computer, they write directions for the on-screen "turtle" to make shapes. They direct the turtle with commands such as fd 50, which means "go forward 50 turtle steps" and rt 90, or "turn right 90 degrees." Off the computer, they solve problems involving shapes, lengths, and turns. They work on problems such as, How much should I turn to make the corner of a triangle with equal sides? If I want the triangle to have a perimeter (the distance around the whole figure) of 60 turtle steps, how long should each side be?

During this unit, you may help your child by:

■ playing the Maze Game with your child when he or she brings it home. Your child can teach you the rules—and probably can let you in on some good strategies. Talk with your child about any other homework assignments as well. Don't solve the problems for your child, but be a good listener and show interest in your child's strategies and questions.

■ talking with your child about paths you walk inside and outside the house. What directions could you give for these paths?

■ listening to your child's plans for his or her "face" design, the final project in the unit. Your child can explain to you how the plan on paper becomes a computer program.

Thank you for your continuing interest in your child's mathematics work.

Sincerely,

Maze Paths

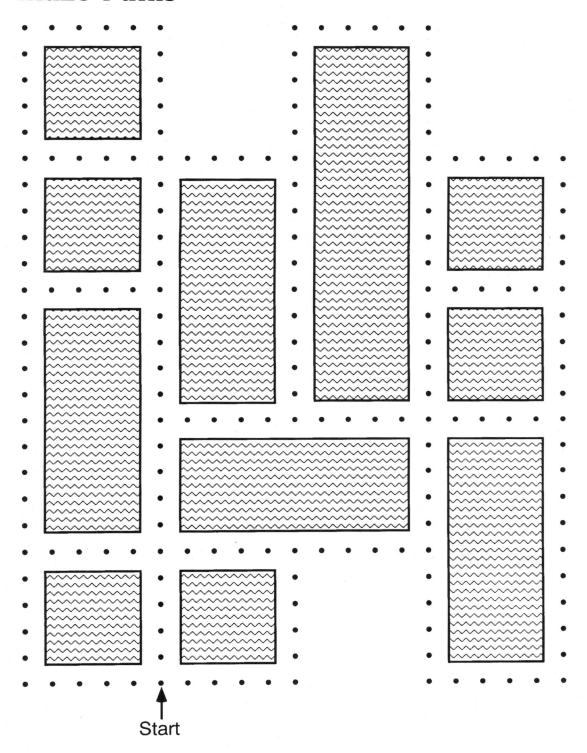

Start

Maze Paths Challenges

Draw each path below in a different color.
Begin each path on the start dot.

Compare your paths with your partner's.
Are they the same?

1. Draw a path 14 steps in length that is open and has 2 corners.

2. Draw a path 40 steps in length that is closed.

3. Draw a path 36 steps in length that crosses itself.

4. Draw a path 42 steps in length with 3 corners.

5. Draw a path 49 steps in length with 9 corners. (A super challenge!)

6. Is it possible to draw a path 15 steps in length that is closed? Why or why not?

Your Maze Path Challenge

Make the longest closed path you can that does not cross itself, beginning on the start dot.

Compare your path with your partner's path. Are they the same path? Is one path longer?

Paths at Home

Choose two paths you often follow at home. For each path, write the starting point and ending point in the blanks below. Then write commands to a robot to follow the path. Use fd (forward), bk (back), rt (right), and lt (left) for your commands. Remember that rt 90 and lt 90 mean turns that look like the corner of a piece of paper. If you need to make turns of different sizes, use a command that describes where the robot would face after turning, such as:

lt to face the windows or rt to face the table.

Path 1

Starting Point _____ Ending Point _____

Commands

_____

_____

_____

_____

Path 2

Starting Point _____ Ending Point _____

Commands

_____

_____

_____

_____

Floors 1 and 2

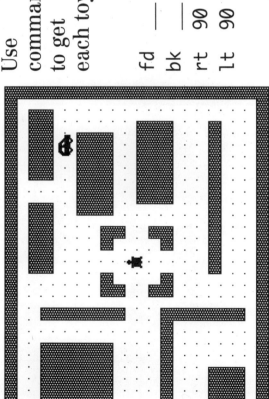

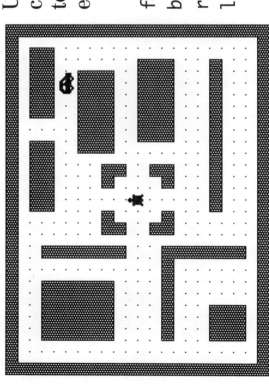

Use
commands
to get
each toy.

fd ___
bk ___
rt ___ 90
lt ___ 90

floor 2

_____

floor 1

Floor 3

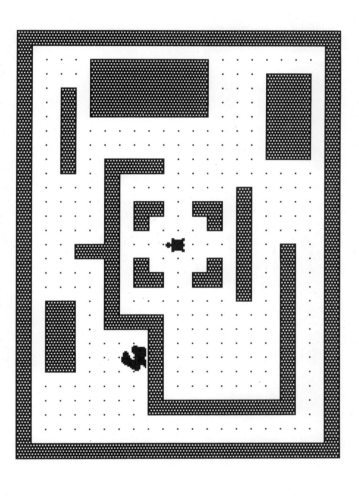

floor 3

Many Possible Paths

Use a copy of Student Sheet 1, Maze Paths, to find and mark the following paths. Draw each path in a different color. You can begin each path at any dot on the page.

1. Draw a closed path 20 steps in length.

2. Draw a different closed path 20 steps in length.

3. Draw another different closed path 20 steps in length.

4. Can you find even more different paths that are like the three you just drew? How many more can you find?

5. Draw a path 25 steps in length with three corners.

6. Draw a different path 25 steps in length with three corners.

7. Draw another different path 25 steps in length with three corners.

8. Can you find even more different paths that are like the three you just drew? How many more can you find?

How to Get Started

- Open *Geo-Logo* with a double-click on the icon.
- Single-click on the opening screen.
- Single-click an activity button.

Geo-Logo Turtle Paths

Geo-Logo Commands

`fd 50`	moves forward 50 turtle steps	`ht`	hides the turtle
`bk 120`	moves back 120 turtle steps	`st`	shows the turtle
`lt 90`	turns left 90 degrees	`setc red`	sets the color to red
`rt 60`	turns right 60 degrees		

Tools

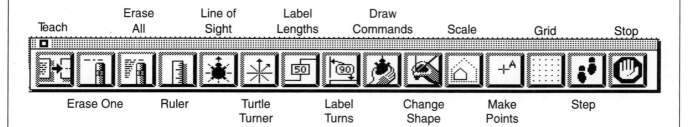

How to Get Help from *Geo-Logo*

- Choose a topic from the **Help** Menu.

How to Open Saved Work

- Turn on the computer, open *Geo-Logo*, select an activity.
- Choose **Open My Work** from the **File** menu.
- Click the name of your work.
- Click .

How to Save Your Work

- Choose **Save My Work** from the **File** menu.
- First time, type a name like **DC+GW toys 3/23.**
- Click ⬚ Save ⬚ .

How to Finish

- Finish activity: Choose **Close My Work** from the **File** menu.
  STOP HERE if changing users.
- Finish *Geo-Logo*: Choose **Quit** from the **File** menu.
- Shut down and turn off the computer.

Turning the Turtle

First, estimate the amount of turn for each path. How much would the turtle have to turn to get ready to draw the dotted part of the path? **Turn your body to help you make your guess.** Write a turn command, such as rt 90 as your guess. Then, use the Turtle Turner to measure the turn and check your guess.

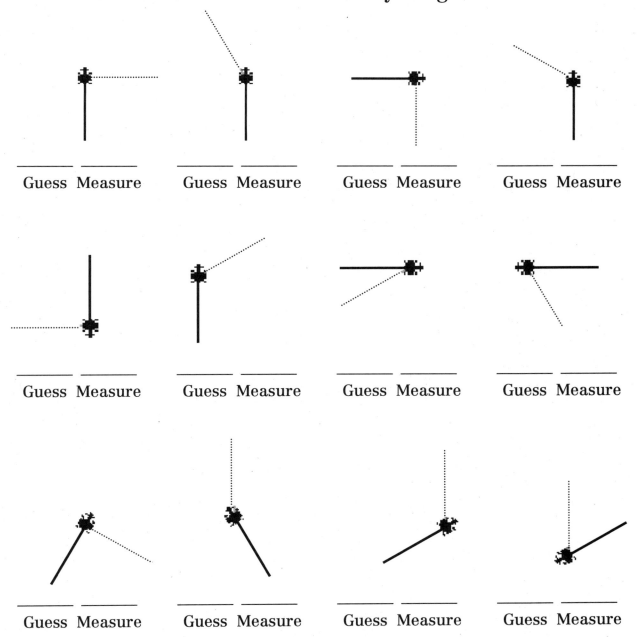

_____ _____
Guess Measure

_____ _____
Guess Measure

_____ _____
Guess Measure

_____ _____
Guess Measure

_____ _____
Guess Measure

_____ _____
Guess Measure

_____ _____
Guess Measure

_____ _____
Guess Measure

_____ _____
Guess Measure

_____ _____
Guess Measure

_____ _____
Guess Measure

_____ _____
Guess Measure

Turn Commands

1. Give three commands for your partner to walk. The turns should always use 30's numbers: 30, 60, 90, 120, 150, or 180.

2. Both you and your partner draw a picture of the path. Use your Turtle Turners.

3. Compare your drawings.

4. Take turns giving the commands.

Example:

Commands	Drawing
fd 3 rt 60 fd 3	

Commands	Drawing

Commands	Drawing

Commands	Drawing

Commands	Drawing

Commands	Drawing

As the World Turns

Estimate the amount of each turn. Then use the Turtle Turner to measure the turn.

	Guess	Measure

1. Bending your pointer finger as far as you can while keeping the joint nearest your fingernail straight. _____ _____

2. Bending your arm at your elbow. _____ _____

3. Turning your head as far as possible from the left to the right. _____ _____

4. Stretching your middle finger apart from your ring finger. _____ _____

5. Opening a pair of scissors. _____ _____

6. Turning your bedroom doorknob just far enough to open the door. _____ _____

7. Opening a jar of peanut butter. _____ _____

Turning of the minute hand of a clock during:

8. A half-hour television show. _____ _____

9. The time it takes to brush your teeth. _____ _____

10. The time you worked on this homework. _____ _____

Feed the Turtle Commands

Use commands
to get to each
pair of berries.

fd __

bk __

rt __

lt __

Cut out and use this
ruler. You will also need
to use a Turtle Turner.

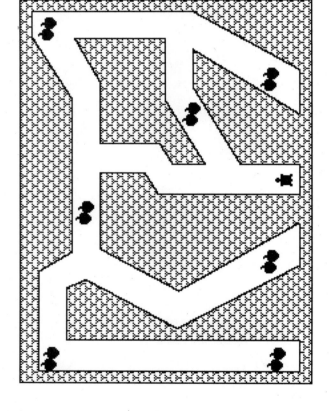

pond

Tricky Triangles

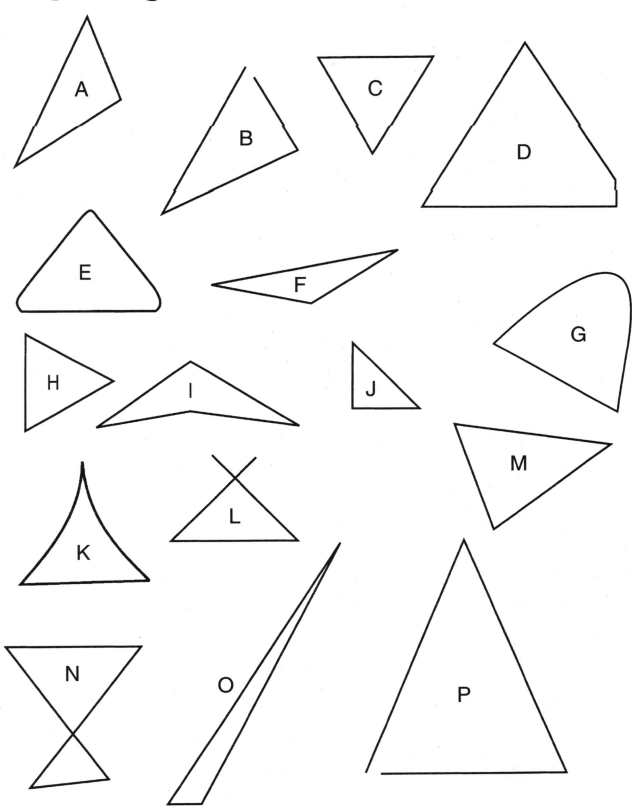

© Dale Seymour Publications®

Investigation 2 • Session 3
Turtle Paths

Which Are Triangles?

Which of the shapes on the Tricky Triangles sheet
are actually triangles?

Secretly, record your answers below.

Discuss your ideas with your partner,
especially those you disagree on.

Write why you think certain shapes are triangles
and others are not triangles.

Triangles	**Not Triangles**

Why?

_____

_____

_____

_____

On the back of this sheet, draw some shapes that are
triangles and some shapes that are not triangles.

Trade sheets with your partner. Put a T in each shape
your partner drew that is a triangle. Put an NT in each
shape your partner drew that is not a triangle.

Triangle Cat

How many triangle paths were drawn to make the cat? _____

Color all the triangles that are exactly the same size
and same shape the same color.

Investigation 2 • Session 3
Turtle Paths

Writing About Triangles

1. Pick one triangle and one shape that is not a triangle. Explain your choices.

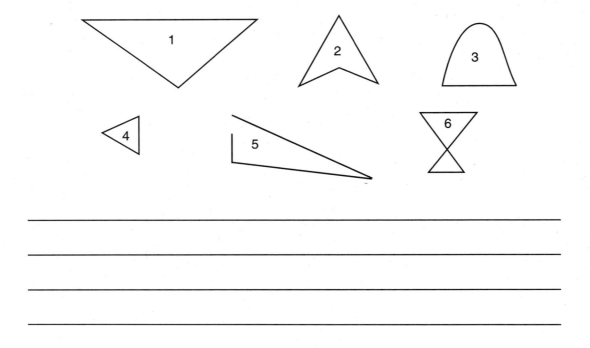

_____

_____

_____

_____

2. Pretend you are talking on the telephone to someone who knows nothing about triangles and nothing about *Geo-Logo*. What exactly would you say to help the person draw a triangle?

_____

_____

_____

3. Pretend you are talking to someone else who does know *Geo-Logo*. What *Geo-Logo* commands could you give to help the person draw an equilateral triangle? Write your commands on the back of this paper.

Missing Lengths and Turns

Finish each figure so it is a closed path. Label the missing
lengths. Then write a procedure for each closed figure.

1. window

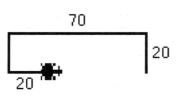

_____ _____

_____ _____

_____ _____

2. door

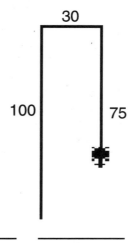

_____ _____

_____ _____

_____ _____

_____ _____

3. flag

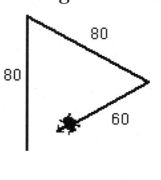

_____ _____

_____ _____

_____ _____

_____ _____

4. factory

20
20
20
20
15

_____ _____

_____ _____

_____ _____

_____ _____

_____ _____

_____ _____

5. steps

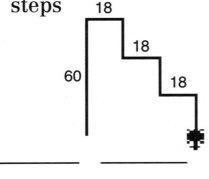

_____ _____

_____ _____

_____ _____

_____ _____

_____ _____

_____ _____

_____ _____

6. house

45   45
45
30 25
40

_____ _____

_____ _____

_____ _____

_____ _____

_____ _____

_____ _____

Investigation 2 • Sessions 5–6
Turtle Paths

Help Make Toys

A toymaker planned to make some toys.
Help her figure out the missing measures
so she can build the toys.

1. Label each line with its length.

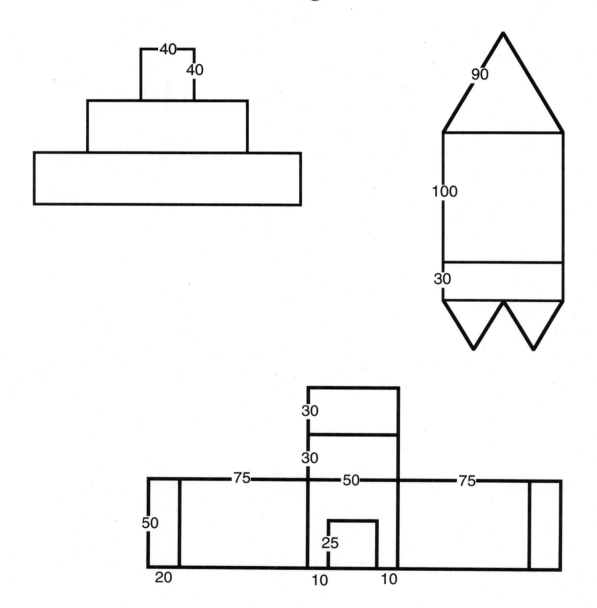

2. Create some toys of your own. Label all the measures.
    Use the back of this paper.

More Missing Measures

Fill in the lines that are labeled with a "?".

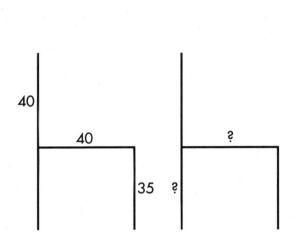

Two identical chairs

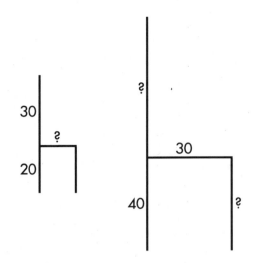

A small chair and a chair that is twice as big

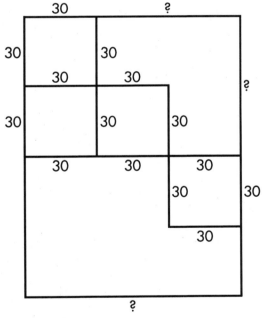

Fancy windows

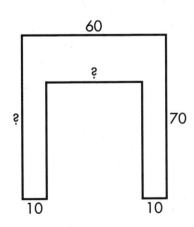

An arch

Make up your own missing measure problem on the back of this sheet. Mark some of the lines with numbers that tell how long they are in turtle steps. Mark some other lines with a "?". Bring your problem back to school tomorrow.

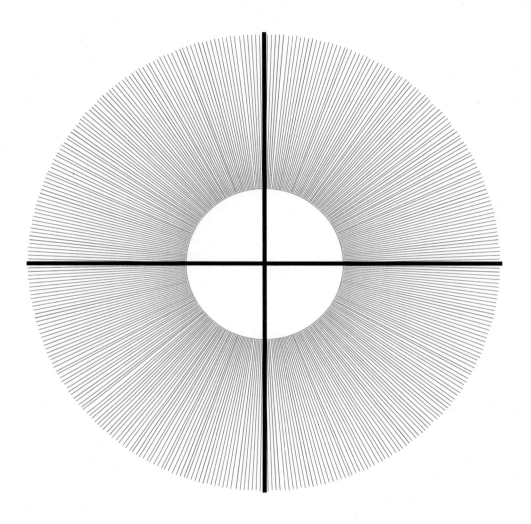

152

TURTLE TURNERS

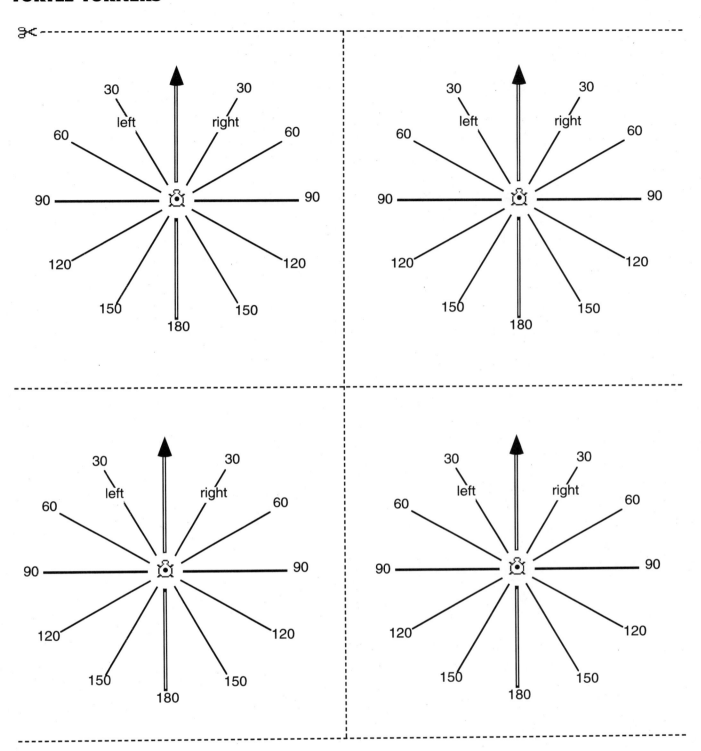

200 Steps and Four 90° Turns

How many different closed paths can you make that have perimeters of 200 turtle steps and four 90° turns? Sketch the paths below, writing the length beside each part of a path.

Are all the figures rectangles? Why or why not?

Challenge: Try to make some rectangles with side lengths that are not multiples of 10—that is, that don't have a zero in the one's place.

200-Step Paths That Are *Not* Rectangles

1. Write commands for a closed or open path that has a total length of 200 steps and more than four 90° turns.
   Can you use this design?
   Can you make another design?
   Check your procedure on the computer in the 200 Steps activity.

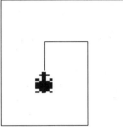

2. Write commands for a closed path with a length of 200 steps that is *not* a rectangle.
   How many turns might you use?
   What size might they be? Check your procedure on the computer in the 200 Steps activity.

3. **Challenge:** Try to write a procedure for a path with a length of 200 steps that does not use 90° turns.

Which Are Rectangles?

Look carefully at each procedure below.
Which ones would form rectangles?
If a procedure does not form a rectangle,
change it so it does. Try to write each procedure using no
more than 8 commands, with the last one being rt 90.

1.	2.	3.
fd 50	fd 65	fd 40
fd 23	rt 90	rt 40
rt 90	fd 65	rt 40
fd 83	rt 90	fd 82
rt 90	fd 23	rt 80
fd 73	fd 42	fd 40
rt 90	rt 90	rt 80
fd 73	fd 50	fd 82
fd 10	fd 15	rt 90
rt 90	rt 90	

Rectangle? _____   Rectangle? _____   Rectangle? _____
   (yes or no)   (yes or no)   (yes or no)

_____   _____   _____

_____   _____   _____

_____   _____   _____

_____   _____   _____

_____   _____   _____

_____   _____   _____

_____   _____   _____

Cutting and Combining Rectangles

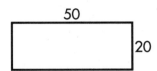

Answer the following questions using the rectangle above:

1. What is the perimeter of this rectangle?

2. What would be the perimeter of a rectangle made by cutting this rectangle in half from top to bottom? Use the picture below to help you figure it out.

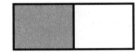

3. What would be the perimeter of a rectangle made by cutting this rectangle in half from left to right? Use the picture below to help you figure it out.

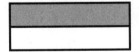

4. What would be the perimeter of a rectangle made by putting two of the first rectangles together, one on top of the other. Use the picture below to help you figure it out.

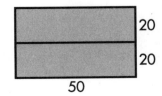

Geo-Face Plans (page 1 of 2)

Work with a partner to plan and make a Geo-Face.
Draw your face design in the head outline on page 2.
Label the line lengths on each of your face parts on the
drawing. Write *Geo-Logo* commands for the face parts
on the back of this paper. When making your face parts,
follow the rules below.

1. All the face parts must be rectangles, squares, or
   triangles.

2. You must have at least one rectangle, one square,
   and one triangle.

3. The perimeter, or total length, of these paths must
   be as follows:

Face Part	Perimeter
nose	90
mouth	200
eyes	100 each
ears	120 each

4. The last command you write for each face part
   procedure should turn the turtle so it is facing up,
   in the direction it started.

5. **Challenge:** Add more face parts:

Face Part	Perimeter
teeth	36 each
eyeballs	40 each

Geo-Face Plans (page 2 of 2)

Perimeter Challenges 1

1. Estimate the perimeter of different objects.
   Add some objects of your own to the list.
   Then measure to check.

Object	Perimeter Guess	Perimeter Measure
a piece of paper		
your desktop		
a book cover		
a table		
your chair		

2. Find two objects with the same perimeter but different shapes. Write about what you found.

3. Measure the height of your chair. Find an object that has a perimeter the same as that measure. Write about what you found.

Perimeter Challenges 2

You need to use six square tiles or squares of paper.
Make arrangements using all six squares.
Full sides of squares must be touching.

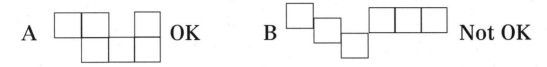

A   OK   B   Not OK

1. Make an arrangement of the squares. Draw it.

   What is the perimeter in side lengths? _____ side lengths
   (Count the number of sides in the perimeter.
   The perimeter of figure A is 14 side lengths.)

2. Make several different arrangements of the squares.
   Draw them below. Write the perimeter, in side lengths,
   beside each drawing.

3. What can you say about the perimeters of different
   arrangements? On the back of this sheet, write
   about what you found.

Practice Pages

This optional section provides homework ideas for teachers who want or need to give more homework than is assigned to accompany the activities in this unit. The problems included here provide additional practice in learning about number relationships and in solving computation and number problems. For number units, you may want to use some of these if your students need more work in these areas or if you want to assign daily homework. For other units, you can use these problems so that students can continue to work on developing number and computation sense while they are focusing on other mathematical content in class. We recommend that you introduce activities in class before assigning related problems for homework.

The Arranging Chairs Puzzle This activity is introduced in the unit *Things That Come in Groups*. If your students are familiar with the activity, you can simply send home the directions so that students can play at home. If your students have not done this activity before, introduce it in class and have students do it once or twice before sending it home. Early in the year, ask students to work with numbers such as 15, 18, and 24. Later in the year, as they become ready to work with larger numbers, they can try numbers such as 32, 42, or 50. You might have students do this activity two times for homework in this unit.

Doubles and Halves This type of problem is introduced in the unit *Mathematical Thinking at Grade 3*. Here you are provided three of these problems for student homework. You can make up other problems in this format, using numbers that are appropriate for your students. Students record their strategies for solving the problems, using numbers, words, or pictures.

Story Problems Story problems at various levels of difficulty are used throughout the *Investigations* curriculum. The three story problem sheets provided here help students review and maintain skills that have already been taught. You can make up other problems in this format, using numbers and contexts that are appropriate for your students. Students solve the problems and then record their strategies, using numbers, words, or pictures.

How Many Legs? This type of problem is introduced in the unit *Things That Come in Groups*. Provided here are two such problems for student homework. You can make up other problems in this format, using numbers that are appropriate for your students. Students record their strategies for solving the problems, using numbers, words, or pictures.

The Arranging Chairs Puzzle

What You Will Need

30 small objects to use as chairs (for example, cubes, blocks, tiles, chips, pennies, buttons)

What to Do

1. Choose a number greater than 30.

2. Figure out all the ways you can arrange that many chairs. Each row must have the same number of chairs. Your arrangements will make rectangles of different sizes.

3. Write down the dimensions of each rectangle you make.

4. Choose another number and start again. Be sure to make a new list of dimensions for each new number.

Example
All the ways to arrange
12 chairs

Dimensions
1 by 12
12 by 1
2 by 6
6 by 2
3 by 4
4 by 3

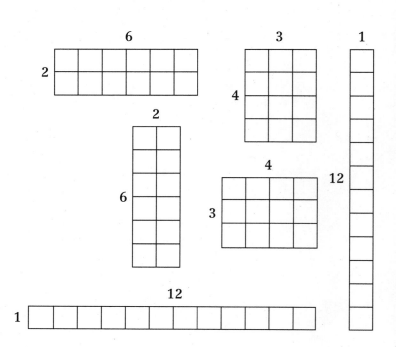

Practice Page A

There are 150 napkins in a bag. They are arranged in two piles. How many napkins are there in each pile?

Show how you solved this problem. You can use numbers, words, or pictures.

Practice Page B

The class has collected 99 empty soda cans to recycle.
They will put them into two large bags. Can they have
the same number of soda cans in each bag?

Show how you solved this problem. You can use
numbers, words, or pictures.

Practice Page C

There are 120 people wearing shoes in the train station. How many shoes are there?

Show how you solved this problem. You can use numbers, words, or pictures.

Practice Page D

Six friends go to the fair together. They pay $18 to get in. How much does it cost each one to get in?

Show how you solved this problem. You can use numbers, words, or pictures.

Practice Page E

Ana and I made lemonade to sell last week. We made three pitchers. Each pitcher made 9 small cups. How many cups of lemonade could we have sold?

Show how you solved this problem. You can use numbers, words, or pictures.

Practice Page F

Jaime has invited 11 boys to his birthday party. His mother will make tacos. She thinks that each boy, including Jaime, will eat 3 tacos. How many tacos will she need to make?

Show how you solved this problem. You can use numbers, words, or pictures.

Practice Page G

Our neighbors, a family of 5, have two hamsters, a cat, and a dog. Cats, dogs, and hamsters have 4 legs. How many legs are there altogether in their house?

Show how you solved this problem. You can use numbers, words, or pictures.

Practice Page H

Show how you solved each problem. You can use numbers, words, or pictures.

There are 32 legs in the stables, and they all belong to horses.

How many horses are there?

How many horses would there be if there were 128 legs in all?